Gender Inclusive Engineering Education

Routledge Research in Education

Manufacturing Citizenship
Education and Nationalism in Europe, South Asia and China
Edited by Véronique Bénéï

Spatial Theories of Education
Policy and Geography Matters
Edited by Kalervo N. Gulson and Colin Symes

Balancing Dilemmas in Assessment and Learning in Contemporary Education
Edited by Anton Havnes and Liz McDowell

Policy Discourses, Gender, and Education
Constructing Women's Status
Elizabeth J. Allan

Improving Teacher Education through Action Research
Edited by Ming-Fai Hui and David L. Grossman

The Politics of Structural Education Reform
Keith A. Nitta

Political Approaches to Educational Administration and Leadership
Edited by Eugenie A. Samier with Adam G. Stanley

Structure and Agency in the Neoliberal University
Edited by Joyce E. Canaan and Wesley Shumar

Postmodern Picturebooks
Play, Parody, and Self-Referentiality
Edited by Lawrence R. Sipe and Sylvia Pantaleo

Play, Creativity and Digital Cultures
Edited By Rebekah Willet, Muriel Robinson and Jackie Marsh

Education and Neoliberal Globalization
Carlos Alberto Torres

Tracking Adult Literacy and Numeracy Skills
Findings in Longitudinal Research
Edited by Stephen Reder and John Bynner

Emergent Computer Literacy
A Developmental Perspective
Helen Mele Robinson

Participatory Learning in the Early Years
Research and Pedagogy
Edited by Donna Berthelsen, Jo Brownlee, and Eva Johansson

International Perspectives on the Goals of Universal Basic and Secondary Education
Edited by Joel E. Cohen and Martin B. Malin

The Journey for Inclusive Education in the Indian Sub-Continent
Mithu Alur and Michael Bach

Traveller, Nomadic and Migrant Education
Edited by Patrick Alan Danaher, Máirín Kenny and Judith Remy Leder

Perspectives on Supported Collaborative Teacher Inquiry
Edited by David Slavit, Tamara Holmlund Nelson and Anne Kennedy

Mathematical Relationships in Education
Identities and Participation
Edited by Laura Black, Heather Mendick and Yvette Solomon

Science, Society and Sustainability
Education and Empowerment for an Uncertain World
Edited by Donald Gray, Laura Colucci-Gray and Elena Camino

The Social Psychology of the Classroom
Elisha Babad

Cross-Cultural Perspectives on Policy and Practice
Decolonizing Community Contexts
Edited by Jennifer Lavia and Michele Moore

Education and Climate Change
Living and Learning in Interesting Times
Edited by Fumiyo Kagawa and David Selby

Education and Poverty in Affluent Countries
Edited by Carlo Raffo, Alan Dyson, Helen Gunter, Dave Hall, Lisa Jones and Afroditi Kalambouka

What's So Important About Music Education?
J. Scott Goble

Educational Transitions
Moving Stories from Around the World
Edited by Divya Jindal-Snape

Globalization, the Nation-State and the Citizen
Dilemmas and Directions for Civics and Citizenship Education
Edited by Alan Reid, Judith Gill and Alan Sears

Collaboration in Education
Edited by Judith J. Slater and Ruth Ravid

Trust and Betrayal in Educational Administration and Leadership
Edited by Eugenie A. Samier and Michèle Schmidt

Teaching and Learning with Technology
Beyond Constructivism
Edited by Concetta M. Stewart, Catherine C. Schifter and Melissa E. Markaridian Selverian

Gender Inclusive Engineering Education
Julie Mills, Mary Ayre, and Judith Gill

Gender Inclusive Engineering Education

Julie Mills, Mary Ayre, and Judith Gill

Routledge
Taylor & Francis Group
New York London

First published 2010
by Routledge
270 Madison Avenue, New York, NY 10016

Simultaneously published in the UK
by Routledge
2 Park Square, Milton Park, Abingdon, Oxon OX14 4RN

Routledge is an imprint of the Taylor & Francis Group, an informa business

Typeset in Sabon by IBT Global.
Printed and bound in the United States of America on acid-free paper by IBT Global.

Library of Congress Cataloging in Publication Data

Mills, Julie.
Gender inclusive engineering education / Julie Mills, Mary Ayre, and Judith Gill.
p. cm.—(Routledge research in education)
Includes bibliographical references and index.
1. Women engineers—Education. 2. Educational equalization. 3. Engineering—Study and teaching (Higher) 4. Sex differences in education. I. Ayre, Mary. II. Gill, Judith, 1942– III. Title.
T65.M5833 2010
620.0071'1—dc22
2009045676

ISBN10: 0-415-80588-0 (hbk)
ISBN10: 0-203-85195-1 (ebk)

ISBN13: 978-0-415-80588-9 (hbk)
ISBN13: 978-0-203-85195-1 (ebk)

Dedicated to Craig, Ken and Peter who supported us in the writing of this and all of our other endeavours and to Geoff, Ian and Jacqui who also supported us and kept the Mills family functioning while Julie was so often at the computer.

Contents

Tables

Foreword by Emeritus Professor Robin King

This new book, *Gender Inclusive Engineering Education*, makes a timely and distinctive contribution to the body of literature concerned with the persistently small numbers of women students in university engineering programs, particularly in English-speaking industrialized countries. The book tackles, in considerable depth, problematic issues raised in numerous reports on engineering education. It specifically addresses findings and proposed actions in the 2007–2008 review of Australian engineering education that I managed on behalf of the Australian Council of Engineering Deans[1]. As a long-time admirer of the authors' individual and collective work, I was delighted to be asked to write this foreword.

The report of that Australian review noted, with regret, the post-2000 demise of most of the earlier initiatives to increase the participation of women, and recorded the rise and then decline in the proportion of women in Australian engineering schools from a peak of 20 per cent to around 16 per cent in 2006. The report called for re-energizing the best of the earlier initiatives, and also commented on the unnecessary persistence of masculine stereotypes within many physical science and engineering topics, and male-oriented approaches to teaching and assessment. Aiming for educational best-practice, the review advocated greater attention to gender inclusive curriculum principles.

By addressing specifically the often neglected area of gender inclusive curriculum—what is taught, and how it is taught and assessed—the authors of this book have provided a very valuable resource. The material is well-grounded, comprehensive and practical. Although it focuses mainly on addressing the interests and aspirations of women students, faculty members and leaders in engineering education will be able to use the material to improve to the quality of engineering education for all students, taking into account the increasing diversity of contemporary student cohorts.

Probably in common with many (male) engineering academics of my generation, my perception and understanding of women's low level of participation in engineering has developed significantly over the past decades. Social and global forces, feminism, the change from elite to mass higher education in which— at least in Australia—women students are now the majority, and women's increased employment opportunities and success in most professions, set contexts for individuals' and organizations' thinking about the low participation of women in engineering study and subsequent careers.

The issue of participation first gained explicit attention in the 1980s, and many of the factors of concern cited at that time persist, admittedly greatly ameliorated by specific initiatives and social changes. The 1995–1996 review of engineering education in Australia[2], for example, urged engineering educators 'to address the serious social imbalance in students seeking to enter . . . engineering . . . particularly the low proportion of women'. For prospective women students and universities, this was then regarded primarily as an equity issue, of increasing women's study opportunities. The report also argued that the academy and profession would benefit 'by recognizing and valuing the alternative outlooks that women . . . may bring to engineering'. This may be much more challenging to achieve.

Alongside the moves to outcomes-based professional accreditation systems that have been implemented in many countries since the 1990s, some of these outlooks were implicitly and sometimes explicitly valued by increasing curriculum attention to the development of graduates' management and communication skills, and to the importance of contextual connections and sustainability in engineering. Such curriculum reform and development, and the increasing adoption of more active learning methods, have benefitted all students. In any event, in some countries, women have a higher graduation completion rate than their male classmates.

Many specific initiatives during the late 1990—mostly in schools' outreach and recruitment—contributed to increasing the proportion of women in engineering schools, but this has slipped back in recent years to under 20 per cent in many countries. This book will support engineering teachers and leaders in engineering education setting out to introduce initiatives to reverse these trends, providing valuable material that will assist those intending to embark on initiatives to increase the participation of women in engineering study.

For engineering faculty members and program leaders, the most immediately valuable parts of this book are likely to be the two chapters on strategies for curriculum design and implementation at the program level and on implementing gender inclusive courses. These chapters are illustrated with examples of making curriculum design, delivery and assessment more inclusive of the interests of women, and indeed other minorities. There are tabulated lists of what curriculum designers should think about, together with practical suggestions for inclusive methodologies. These comprehensively researched and referenced examples and lists will provide engineering academics with material for thinking about their teaching as a whole, rather than (merely) increasing its inclusivity. Academics concerned with ensuring their programs meet the graduate attributes typical of professional accreditation systems will find value in the strategy table on improving inclusivity on curricula, teaching and assessment. The final 'practical' chapter on embedding gender inclusive practices rightly stresses the importance of institutional policies and visible leadership. Deans should not underestimate the value of special support initiatives for women faculty.

These practical 'how to do it' chapters are preceded by material that sets the context and argues the case for the gender inclusive curriculum. The summary list of principles that places respect for the individual student and commitment to their potential (albeit in group educational situations) at the core of the education process may seem obvious, but it is a valuable reminder to all educators of their purpose. Many readers will find the comparative national data on the participation of women in engineering education to be an informative reference against which the outcomes of actions in many of the countries cited can be tracked. The chapter on initiatives to attract more women into engineering will provide readers committed to developing and promoting new initiatives, as mentioned earlier, with useful insights.

The chapter on the theoretical contexts of learning, curriculum and gender will be of great value to the majority of engineering academics who do not have formal qualifications in teaching. Such readers will learn that educational constructivism aligns with many of the principles of engineering, yet may also reflect on the fact that their teaching of basic engineering material may not truly embody these principles. Readers will also be brought face-to-face with the notions of student identity and self-concept in education, and their link to curriculum. This chapter would provide any engineering program leader with material to challenge those colleagues who may maintain that engineering education is somehow gender neutral.

In summary, readers of this book who take into their engineering education practice the authors' knowledge, experience and insights contained in this book will surely be contributing positively to the quality of the education for all of their students. As leading educators, they will be contributing directly and indirectly to increasing the participation of women and other minorities, and thereby enhancing the quality of engineering, a key profession for tackling many of the future challenges humanity faces. They will be doing what so many of our prospective engineering students say they themselves want to do: 'to make a difference'.

Robin King
Professor Emeritus,
University of South Australia

NOTES

1. King, R. (2008), *Engineers for the Future: addressing the supply and quality of Australian engineering graduates for the 21st century*, Australian Council of Engineering Deans (www.aced.edu.au).
2. IEAust (1996), *Changing the Culture: engineering education into the future*, Institution of Engineers Australia, Canberra.

Acknowledgments

We would like to acknowledge the support of the grant we obtained from the Australian Learning and Teaching Council Ltd, an initiative of the Australian Government Department of Education, Employment and Workplace Relations in providing partial funding towards this project. In particular we would like to thank our research assistant Ms Lee Clark for her careful editing and assistance in producing the final manuscript and Mr Jeff Allen for his assistance in sourcing statistics for Chapter 2.

1 Introduction

Why We Need Gender Inclusive Engineering Education

> Few attempts have been made to revise the engineering curriculum itself to promote diversity, even though it is clear that the curriculum is central to what defines an engineering education. Instead most diversity initiatives aimed at the undergraduate engineering student population have started with a curriculum that is known to be unattractive to women and minorities . . .
>
> (Busch-Vishniac and Jarosz 2004: 256)

THE ELEPHANT IN THE CLASSROOM?

Despite many campaigns and initiatives over the past 30 years, in western economies white males continue to dominate numerically in both the engineering profession and the students seeking entry to it. Many otherwise excellent books and reports on engineering education continue to appear, which do not acknowledge this problem. Gender thus remains the elephant in the engineering classroom—the knowledge of its presence is widely shared, its shape is unmistakeable but no one talks about it.

In this book we propose to talk about the 'Who?' question as well as the 'What?' and 'How?' of engineering education because we believe that the composition of the student body is an important component of the teaching and learning process. We argue that curriculum involves much more than the set of technical skills and capacities expected of engineers. A National Science Foundation report observes that the engineering curriculum and the culture it creates are 'at odds with the value systems of most young women and minorities, and . . . probably at odds with many talented students of any race and gender' (NSF 2005: 36). Today's engineering students come from different racial, cultural, educational and language backgrounds; many of them are male and some are female. Our immediate aim is to alert engineering educators to the ways in which these differences in the student body impact on the teaching and learning in engineering classrooms. Ideally educators will use these differences to create an improved learning context for all student engineers. The long-term goal is to produce better engineering education and consequently more entrants to this important profession.

THE IMPORTANCE OF ENGINEERING

Engineering is essential for the sustainable development of modern society. In the developed world, engineers create and implement technologies that are critical for maintaining and improving the quality of life. Their products include advanced communication systems, environmentally sustainable buildings, complex transport networks, sanitation and water supply systems as well as the infrastructure, technologies and materials used in industries which are vital to the global economies. In the less-developed world, engineers improve living conditions by adapting these same technologies and industries so that they are appropriate for use in areas where they are currently non-existent or only at a basic level. Despite the widely recognized value of engineering and the urgent need for more engineers, many countries face difficulties in finding graduates with the necessary skills. The number of tertiary students of engineering has been declining for the past two decades in many western countries. This situation is further exacerbated by the decreasing number of high school students graduating with mathematics and science at a level regarded as acceptable to enter engineering programs. Poor retention of engineering students at tertiary level and of professional engineers within the workforce is also a critical issue. As a consequence of all of these factors, engineering faces even more difficulty in overcoming current and predicted skills shortages when compared with other professions. 'Engineering has never mattered more'. So begins a recent book on engineering education (Sheppard et al. 2009). The authors build a most convincing argument about the importance of engineering in terms of its ubiquity and relevance to today's world. However, they do this without any reference to the student body in terms of its diversity or the gender issues in its makeup. While endorsing the sentiments about the importance of engineering and of making engineering education a more engaging future-oriented practice, this book directly addresses the ways in which engineering curriculum needs to change to be more inclusive of the diversity of the student body, especially in terms of gender. That is, it recognizes the elephant in the room.

A MALE DOMINATED PROFESSION

Notwithstanding the looming shortages, engineering stands out as one profession in the western world that remains overwhelmingly male dominated, both within its student ranks and even more so in the engineering workforce. In most western countries, the domination is not only male but white male—a 'pale male profession' as described by William A. Wulf, a former president of the US National Academy of Engineering (Cornell Chronicle 2001, cited in Galloway 2008: 52). To meet industry's demand for engineering graduates, it is critical that universities draw from the total pool of qualified students available, not just the white, male proportion of it. It would seem obvious that a major step forward in this direction would be to increase the number of females in the engineering classrooms. However, this solution is rarely given the attention or resources needed to make it happen. In many cases women appear to remain invisible to those implementing programs to address skills shortages in these industries. A clear demonstration of this routine oversight appeared in an audit of science, engineering and technology skills conducted by the Australian government in 2006 (DEST 2006), which, among all of its 76 pages detailing the current and future skills shortages in those sectors and the recommendations for addressing them, never once used the word 'women' or 'female', nor presented any gender-disaggregated data, or in any way mentioned the absence of women as an issue to be addressed or from which to develop a potential solution. Fortunately this omission has since been acknowledged and some remedies proposed in a report produced by the Australian Council of Engineering Deans (King 2008).

BEYOND THE BUSINESS CASE FOR MORE DIVERSITY IN ENGINEERING

Whilst the skills shortage is certainly a major concern and can function as a key economic driver to address the issue of gender imbalance in engineering, it is clearly not the only reason for looking at the situation. Engineering needs diversity as much as any other profession and to achieve this it must become more inclusive. As the chief technology officer in Sun Microsystems observed in 2006:

> In the end, society is influenced by a variety of motivations. Whether yours is social justice, personal experience or the old-fashioned profit motive, there is no doubt that being inclusive in engineering will make the whole field richer, wealthier and more connected to society.
>
> (Papadopoulos 2006)

Increasing the number of women in engineering is an issue of social justice, as it directly relates to ensuring that a wider range of citizens play an active and informed part in the control and use of social assets. If, as we have argued, engineers are responsible for decisions and developments that impact on the daily lives of all members of society, then clearly those decisions will be more representative of the desires and needs of that society if the professional workforce that makes them is also representative of that society.

> Approaches and perspectives in engineering have come from a homogeneous group for far too long. If we begin to see how diversity will benefit the research community, the profession, and our countries, then women will finally take their rightful place in the ranks. We will all be the richer for it.
>
> (Frize 1999)

The issue of quality is also crucial in engineering: to ensure that the best and most able people from all backgrounds are provided with the necessary education to contribute to the further development of knowledge. Thus the rationale for increasing the numbers of women in engineering derives from a range of agendas, all consistent with an appreciation of the key role played by engineering in the developed and developing world. Key questions immediately arise: how do we do this and why hasn't it already happened?

THE CURRENT SITUATION

Although the numbers of female students commencing undergraduate engineering studies have increased in most western countries during the past 25 years, they remain low. In addition, the indications are that these numbers have now reached a plateau or are actually in decline in countries such as Australia, Canada, the US and the UK (Gibbons 2006; CCWESTT 2006; UKRC 2006; King 2008). The representation of other minority groups is similarly low. Moreover, recruiting female students to engineering programs is not enough to improve female representation. Once enrolled, these students must remain motivated to succeed in their studies and subsequently to join, and stay in, the profession. Numerous national and international reports and studies have repeatedly shown that there are significant problems in both recruitment and retention of women in engineering (Cronin and Roger 1999; CAWMSET 2000; ETAN 2000; Thom 2001; Greenfield et al. 2002; Blickenstaff 2005; Mills, Mehrtens et al. 2008). While a recent large US study (Ohland et al. 2008) has challenged this picture by demonstrating similar retention patterns for males and females, these rates are still very low (53 per cent), suggesting that the recruitment is not effective and the experience of engineering education does not fit with the aspirations of a significant number of students. Studies report that once women have graduated, many more of them leave the profession than men, a point discussed further in Chapter 2. The challenge faced by the profession is a dual one: to acknowledge the interests and backgrounds of women during their engineering education while simultaneously changing the culture of the engineering workforce and workplace to ensure that women can be comfortable and succeed in their profession (Khazanet 1996; Lewis et al. 2007; Mattis 2007; Gill, Mills et al. 2008).

GENDER AND EDUCATION

The traditional explanation for the lack of women engineers was that, not having studied mathematics and sciences in senior school, women did not qualify for entry. This story has changed significantly in recent times. In the closing decades of the twentieth century, following much interest in girls and education, the long-standing tendency of males predominating in enrolments in these courses, and also more importantly their dominance in the lists of high achievers, had begun to change. Increasing numbers of girls were enrolling in science and mathematics courses in the senior school years, and more of them had started to appear among the high achievers in these subjects (for example Huang et al. 2000; Doherty 2002). This outcome led to a widespread identification of a crisis in boys' education (Epstein 1997; Gorard et al. 1999; Mills 2000; Martino et al. 2004). Across the UK, US and Australia, government-funded committees were set up to solve the problem of male educational underachievement. There was much resourcing of measures aimed at improving the educational outcomes of boys. While several writers insisted that the 'problem' was not particularly new and that middle class boys continued to do as well in school as they had always done (Gorard et al. 1999; Teese and Polesel 2003; Gill 2004), the furore around the idea of a male crisis continued unabated (Paechter 1998; Reed 1999; Gill and Starr 2001). Although the perceived female dominance in end of school examinations was not followed by their concomitant uptake in engineering enrolments, the educational community had accepted the idea that gender equity mattered in educational provision and treatments. However, from the mid 1990s the concerns about boys' education deflected attention from the girls-into-engineering initiatives that had been relatively widespread in the 1980s to 1990s (Phipps 2008). Once again girls and women had disappeared from the agenda of educational reform.

A NEW FOCUS

One aim of this book is to refocus attention on issues in education that are pertinent to women in engineering. Thus we will provide a review and appraisal of the theory and practice of gender inclusive engineering education at the tertiary education level and make practical suggestions as to how the concept can be applied at the program, course and classroom level (see Appendix B for a glossary of terms used). The book incorporates research perspectives on controversial issues relating to women and education and offers new interpretations of an inclusive curriculum. We argue that such a curriculum is necessary if we are to improve the access, retention and success of women in engineering education. We aim to encourage administrators and faculty to engage with a gender inclusive engineering education by providing, along with the surrounding conceptual and practical debates, the rationale for its adoption and practical examples of its application. Furthermore we base our argument on the standpoint that all students, not only female ones, benefit from approaches to admissions procedures, curriculum content, and teaching, learning and assessment methods which acknowledge and include social and cultural differences in students' interests and perspectives. Ultimately we suggest that these approaches are necessary for the transformation of the engineering profession in ways that render it more flexible, inclusive and appropriate for social and global realities.

THE ENGINEERING PROFESSION IN THE CHANGING GLOBAL CONTEXT

The story of engineering forms a continuous narrative of change and adaptation consequent upon broad changes in social organization and scientific progress. Currently the rate of change in the engineering profession is increasing rapidly and the applications and priorities of the profession are changing as a direct consequence of globalization. Issues such as climate change and global warming, mass communication, the supply of potable water, environmental degradation, global health and so on are becoming increasingly critical. In all of these areas, engineering knowledge and problem-solving skills are critically important. As noted by Galloway (2008):

> In the 21st century, advances in engineering will, to a large measure, determine the rate of world economic growth, the quality of life, and standards of health and security. . . . The mission of the 21st century engineer is to ensure that there are no boundaries in terms of how engineering is applied to better the quality of life worldwide.
>
> (Galloway 2008: 12–13)

At the same time, there is increasing competition for engineering goods and services in the global market. Engineers today are increasingly required to be multi-disciplinary and to apply social and environmental perspectives to their technological expertise. They need appropriate communication skills to engage in debates on these issues with technical and non-technical people. In order to remain relevant to the profession it serves, engineering education must also undergo processes of review and change as it addresses these new areas and reforms the more established ones.

Professional organizations world-wide have attempted to profile the engineer of the future including the necessary skills and abilities. Major reviews of engineering education have pointed to the need for change to develop these future engineers (ASEE 1994; IEAust 1996a; NAE 2004, 2005; RAE 2006, 2007; King 2008). These reviews have involved engineering educators along with industry leaders, think tanks, governments and professional engineering organizations, demonstrating a widespread agreement about the need for change in the profession and in the education leading to entry to it.

In the United States, the American Society for Engineering Education (ASEE) in its 1994 report *Engineering Education for a Changing World* stated:

> Today, engineering colleges must not only provide their graduates with intellectual development and superb technical capabilities, but following industry's lead, those colleges must educate their students to work

> as part of teams, communicate well, and understand the economic, social, environmental and international context of their professional activities.
>
> (ASEE 1994)

The more recent US report by the National Academy of Engineering (NAE 2005) updated this assessment and asserted that:

> While certain basics of engineering will not change, the explosion of knowledge, the global economy and the way engineers will work will reflect an on-going evolution that began to gain momentum a decade ago. The economy in which we will work will be strongly influenced by the global marketplace for engineering services, evidenced by the outsourcing of engineering jobs, a growing need for interdisciplinary and system-based approaches, demands for new paradigms of customization, and an increasingly international talent pool. The steady integration of technology in our public infrastructures and lives will call for more involvement by engineers in the setting of public policy and in participation in the civic arena.
>
> (NAE 2005:3)

This position was echoed by the review of the Royal Academy of Engineering in the UK in its 2007 report:

> Today, business environments increasingly require engineers who can design and deliver to customers not merely isolated products but complete solutions involving complex integrated systems. Increasingly they also demand the ability to work in globally dispersed teams across different time zones and cultures.
>
> (RAE 2007: 5)

All of these reviews and reports have emphasized the issues that arise from the increasing globalization of the engineering industry and have linked this to a need for an increased diversity in the engineering workforce. The business imperative for increasing diversity in the workplace has received mounting attention over the past decade. An article in the *Harvard Business Review* neatly summarizes this business case:

> How well an enterprise works—how productive and successful it is in a highly competitive global economy—depends on whether it has the best people and people who are comfortable working across lines of race, class, religion, and background. The days of insularity and parochialism are gone. Diversity is a business imperative because it affects competitiveness.
>
> (Bowen et al. 1999: 140)

The repeated use of terms such as 'global marketplace', 'business environments' and 'highly competitive global economy' indicates the imperative to satisfy the needs of the economy—both national and global. Moreover there is a clear connection between having more of the 'best possible people' in engineering and success in the industry. In addition, evidence of the value of having a gender-balanced workplace comes from the Catalyst (2004) study, which established a quantitative link between the percentage of women in senior management and the financial performance of 353 Fortune 500 companies between 1996 and 2000. The group of companies with the highest representation of women in top management teams had a 35.1 per cent higher return on equity and 34 per cent higher total return to shareholders than the companies with the lowest representation.

In 1996 a major report on engineering education in Australia entitled *Changing the Culture: Engineering Education in the Future* (IEAust 1996a) emphasized the importance of improving the participation of women in engineering education and established a task force to advise on this matter (IEAust 1996b). Among the recommendations were the importance of establishing an inclusive culture and the need to 'create inclusive educational programs that cater for the diverse range of interests, skills and abilities to enrich the future engineering profession . . . ' (IEAust 1996a:152). The release of these reports resulted in some significant changes to engineering education in Australia, particularly in relation to engineering curricula and pedagogy. However, whilst some progress has been made, a recent review of the report outcomes (King 2008) has found it necessary to repeat some of the 1996 recommendations, as well as noting some areas that need further attention. The issue of diversity is raised again, and it is recommended that the engineering profession in Australia should 'address shortages by increasing diversity in engineering workplaces supported by engineering education programs.' (King 2008: 111). One of the means suggested to do this is to 'Define and implement inclusive curriculum for engineering: reducing male stereotypes within the curriculum, and revitalising the best of the Women in Engineering programs' (King, 2008: 116). This work will be the focus of this book.

THE CURRENT SITUATION: A SCARCITY OF WOMEN ENGINEERS

The low representation of women in engineering education and the profession sends a message that the profession cannot afford to ignore. More detailed statistics are provided in Chapter 2, but across most of Europe, North America, Australasia, Africa and Asia the percentage of women currently enrolled to study engineering at institutions of higher education remains below 25 per cent. This figure is often cited as a 'tipping point' for minority status, as critical mass research indicates that below this level students are more likely to discontinue their studies (Kanter 1977; Blum 2001; Powell et al. 2006). In the professional workforce the percentage of women in engineering is significantly lower, often less than 15 per cent, and in many individual workplaces, women engineers may feel very isolated (Gill, Mills et al. 2008).

Some engineering specializations have higher percentages of female students. For example in chemical and environmental engineering the percentage of female students exceeds 25 per cent and sometimes approaches 50 per cent, even in western countries. In the US in 2005–2006, 36 per cent of chemical engineering and 44.2 per cent of environmental engineering bachelor degrees were awarded to women. However, the total number of degrees awarded in these fields represented only 6 per cent and 0.6 per cent respectively of all engineering bachelor degrees awarded in the US that year (Gibbons 2006); hence in overall terms this does little to alter the situation for engineering as a whole. In addition these fields still suffer from the same issues of poor retention once women enter the engineering workforce, so that the percentage of female chemical engineers in the workforce remains below 25 per cent (for example 19.7 per cent in the US in 2003 (Table H-5, NSF 2007) and 18 per cent in Australia in 2007 (Mills et al. 2007)).

In a few countries and regions the numbers of women in engineering education and the professional workforce are significantly higher than elsewhere. They include some Eastern European countries, others that were part of the former Soviet Union and some countries in Latin America. These exceptions and the possible explanations for them will be discussed in more detail in Chapter 2.

So we can conclude that the minority female proportion differs across countries and across specializations within the discipline of engineering. What these data demonstrate is that engineering, like other professions, operates in a social context, and changes in the social context will likely affect the gender balance. This perception opens the possibility for positive interventions in the context of engineering education which can make a difference to gender balance.

PAST EFFORTS TO ATTRACT WOMEN TO ENGINEERING

In recent decades government agencies, industry and educational institutions have expended significant amounts of time and money in attempts to increase the numbers of women enrolling in engineering degrees. However, as noted earlier, the gains have been slim. Frequently the increases in female enrolment that have occurred are not sustained when the funding runs out, or when the key people involved in an initiative are no longer able to continue (Urban Institute Education Policy Center 2000; Clewell and Campbell 2002; NSF 2003; Lewis et al. 2007). There have also been programs directed at primary and secondary education to attempt to interest all students in engineering and the skills of mathematics and science. Some of these initiatives have been particularly focussed on girls and women (Thom

2001; NSF 2003). Again the success has been limited and often short lived in terms of leading to enrolments in engineering.

Attracting students to engineering is only a partial solution, because of the high attrition rate in engineering courses for both women and men (Seymour and Hewitt 1997; Thom 2001; King 2008). For those women who graduate with an engineering degree and go on to employment in the engineering profession (some gain their engineering degree but enter other fields), the prospect is of a workplace where they are likely to get lower pay, be less satisfied with their opportunities for promotion and may face discrimination and possibly harassment on the basis of their gender. Not surprisingly, there is a high attrition rate of women in the professional engineering workforce, with significant numbers leaving within 10 years of entering the workforce (Ambrose et al. 1997; Roberts and Ayre, 2002; CAWMSET 2000; Greenfield et al. 2002; Mills, Mehrtens et al. 2008)

The failure to attract and retain women in engineering has been frequently ascribed to the culture of engineering at university and in the workplace. This culture, it is argued, is male dominated, authoritarian and conservative, described by Ambrose et al. (1997) as follows:

> The culture of science was evolved largely by able-bodied heterosexual white men, and people who do not fit this mold may encounter discrimination ranging from the subtle to the overt: "Outsiders" may not be able to integrate easily with—or may simply dislike—the dominant culture.
> (Ambrose et al. 1997: 24)

There is no evidence of significant change from this position over 10 years later (Gill, Mills et al. 2008). These issues will be discussed in more detail in Chapters 2 to 4. Whilst it is essential to address the issues of women in engineering across the full spectrum of their experience, from early socialization through their school and undergraduate education, early workforce experiences and long-term career outcomes, the primary focus of this book is on the undergraduate education experience and what can be done to improve this for women. Undergraduate education is the 'gatekeeper' for professional careers in engineering and hence it is a critical site for intervention (Davis et al. 1996).

GENDER INCLUSIVE ENGINEERING EDUCATION

Gender profile statistics for women and other minority groups in engineering (see Chapter 2) suggest that the traditional orientation of established engineering educational practice has remained largely unchanged (Gill et al. 2005; Gill, Sharp et al. 2008; Metz 2007). Thus the gender disparity in engineering enrolments is unlikely to diminish unless there is significant change in course promotion and experience. Many programs to increase women's enrolments, while moderately successful, have been short lived and tied to short-term funding and specific champions. They have not evolved into strategies for long-term change embedded within the curriculum. This is not to say that these initiatives are not important or to argue that they should not continue, but the issue of embedding change is critical for sustained improvement.

Gender inclusive engineering education at the college or university level includes a wide spectrum of programs, initiatives and practices, all of which are important components of the strategy to increase the number of women entering and successfully completing engineering undergraduate education and going on to fulfilling professional careers. These include:

- high school outreach programs to attract girls to study engineering
- support and mentoring programs for women enrolled in engineering
- role modelling from senior female students, female faculty members and female professionals from industry
- gender inclusive curriculum that is embedded throughout the program of study
- professional development and training for faculty to develop and deliver gender inclusive curriculum and to support women students
- management level leadership and support for all of the preceding components
- professional and accreditation organization support for all of these components.

These components will be discussed fully in later chapters, but here we reflect on the importance of the curriculum component of this education spectrum, which has been blamed for the difficulties in recruiting and retaining female engineering students. Beder (1989) described engineering curriculum as showing an 'obsession with the technical, the mathematical, and the scientific, and an almost complete neglect of the social, political and environmental issues' which discourages:

> . . . students with broader interests, a different range of talents . . . ; those who want to work with people rather than machines and numbers, those who care about social relations. Too often it is the female students who are put off.
>
> (Beder 1989: 173)

A further illustration of the way in which the engineering curriculum can alienate women is provided by Nair and Majetich (1995):

> Examples drawn from predominantly male sports and from weapon deployment dominate mechanics courses and ignore the backgrounds of female students as they attempt to understand or visualise situations.
>
> (Nair and Majetich 1995: 36)

This curriculum has a pronounced effect on the educational ethos in undergraduate engineering and yet it has been virtually neglected in women in engineering initiatives to date. Davis and Rosser (1996) examined the programs that had been funded by the NSF to increase the presence of women in science, mathematics and engineering in the US in the years to 1994, along with the projects funded by the Women's Educational Equity Act Program and noted that:

> In sum, most of the curriculum reform efforts in science, mathematics, and engineering have not focused on women, and *most of the efforts centered on women in science have not focused on curriculum or classroom issues.*
>
> (Davis and Rosser, emphasis added, 1996: 243)

In 2003 the NSF issued a report on 211 of the total of over 250 grants that had been funded under their Program for Women and Girls scheme analyzed earlier by Davis and Rosser (NSF 2003). The total funding allocated to these programs was over $US90 million since 1993. However, the index lists only 20 of these (less than 10 per cent) under 'curriculum', and closer examination reveals that only seven of these relate to undergraduate curriculum, of which two are directly involved with engineering curriculum. These statistics reiterate our point that the curriculum component of gender inclusive engineering education is clearly in need of further attention.

THE GENDER INCLUSIVE ENGINEERING CURRICULUM

We use the term 'curriculum' throughout this book in its broadest sense and fully define this in Chapter 4. Meanwhile: *content plus teaching and learning methods* may serve as a working definition or, for North American readers, *curriculum plus pedagogy.*

The curriculum provides students with the knowledge and skills required for entry to the engineering profession. The selection of the knowledge and skills to be included in any curriculum is the responsibility of the academic staff or faculty, in conjunction with the national accreditation body, and the selections made will therefore shape student perceptions of the current priorities, attitudes and practices in the profession. The curriculum will not only be the main influence on students' learning experiences, it will also provide the foundation for students' professional lives, influencing their attitudes to creativity, innovation and other change, such as the value of women engineers in the workforce.

For some time, the undergraduate engineering curriculum has been identified as typically catering to a narrow range of student interest and prior experience. In 1995 Lewis (1995) described the curriculum framework in engineering as follows:

> The research questions, methods, criteria of success, and styles of teaching are male defined and consequently, the knowledge itself reflects a bias towards a male cognitive style in its practices, theories and ways of teaching. The science and engineering makers have created disciplines where they are comfortable with the separation of theory from social or environmental context, and where the top down, expert scientist authority is the dominant paradigm.
>
> (Lewis 1995: 270)

And more recently King (2008) noted that:

> Nevertheless, students and others have observed that engineering curricula (and physical science texts) tend to be crafted with over-use of masculine stereotypes and examples, such as automobiles, rockets and weapons.
>
> (King 2008: 72)

The identification of male bias, coupled with the apparent resistance to change in the face of a pressing need for more engineers, leads to the proposal that the traditional orientation of established engineering education curriculum must become 'inclusive' by taking more account of the interests and needs of individuals from all groups in our society (Mills and Ayre 2003; Busch-Vishniac and Jarosz 2004; King 2008). The emphasis here is on the desirability of structuring an engineering curriculum around a general recognition that students from diverse backgrounds bring different perspectives, attitudes and values to the engineering classroom. This can be done without making distinctions between the specific cultural groups or genders represented in the class.

We will develop the principles of an inclusive curriculum in more detail in later chapters, but in summary they are:

- respect and concern for every student as an individual
- recognizing that students are simultaneously individuals and group members and their groups may be differently positioned with respect to the knowledge
- commitment to enabling each student to realize his or her full potential
- awareness that differences exist between individual students in interests, values, perspectives, prior experiences, ambitions, learning styles and home circumstances
- commitment to acknowledging, recognizing, respecting and accommodating these differences in all components of the curriculum
- using, in a professional way, the diversities which exist amongst the students, to enhance the learning experiences of all students
- awareness of, and willingness to address, unstated and possibly unconscious attitudes, values and assumptions in ourselves which impact on student learning environments

- awareness of student-student interactions in the class and a commitment to influencing these if they impact negatively on any of the other principles
- listening to student comments and evaluations of the informal learning environment as well as the formal parts of the course, and taking these into account in reviewing the curriculum.

Men and women from cultures other than those which have developed in the western scientific tradition can also feel alienated, uncomfortable and unrecognized in a system which practises this tradition. By combining gender and culturally inclusive principles—that is, emphasizing recognition of the different values, perspectives and learning styles of all students, including those of the dominant group—an inclusive curriculum can and should be *both* gender and culturally inclusive. These concepts are not mutually exclusive: the principles of a gender and culturally inclusive curriculum are similar and also apply to curricula which are inclusive of other differences between groups and individuals, such as ethnicity, age, disability and sexual orientation.

WHY SO RESISTANT TO CHANGE?

A contributing factor to the resistance to change by engineering educators relates to the degree to which features of the knowledge associated with the discipline are regarded by practitioners as knowledge pure and simple, existing in its own right. However, current learning theory suggests that all knowledge operates in a social environment such that certain sorts of knowledge become associated with times, places and gender. In this way engineering knowledge can be seen as coded masculine, as it has operated for so long in a space largely inhabited by males (Rosser 1995). This point will be developed further in Chapter 4.

An inclusive curriculum works by incorporating a range of teaching, learning and assessment approaches. A range of learning theorists, including Fox Keller (1985), Lewis (1995) and Rosser (1995), suggest that gender differences in approaches to learning arise from a preference amongst women for 'connected' knowing (connecting the self to knowledge) in contrast with the 'separate' critical and analytical approach usually so highly valued in higher education. Rosser, for instance, remarks:

> . . . females are more likely to feel comfortable in approaching problems and laboratory experiments if they understand the relationship of the particular problem or experiment to the broader context of the bigger problem of which this solution may be a small part.
>
> (Rosser 1995: 6)

While we do not want to imply any degree of essentialism, a modification of Rosser's statement which notes that more women than men feel the need for this type of connection may be closer to the mark. We do want to endorse the notion that the use of different approaches in teaching and learning enables all learners to have a better chance to relate to, and to understand, the point at issue.

In the communications revolution of recent decades there has been a significant increase of public awareness of the social impact of engineering applications, nowhere more strongly than in the environmental lobby. We suggest that gender typical differences produced by socialization practices may orient women towards social connection and men towards more abstract functionalist principles (Gilligan 1982; Brew 1999), and there is some indication of support for this thesis in women's higher take up of engineering in those areas that have clear social benefit implications, such as environmental engineering.

Traditionally the role of professional degrees has been to produce people who will fit with the existing professional culture. In contradistinction the case here is being made for universities to require their engineering graduates to demonstrate the necessary knowledge and skills to become change agents within traditional engineering culture, in line with the argument developed earlier. What is needed then is a curriculum which will assist in the production of new-style engineers.

A gender inclusive curriculum can be introduced without loss of rigour in the program: this approach should also improve the quality of the education provided. International studies have clearly shown that inclusive curriculum strategies have improved engagement for all students, not just women, along with significantly improved retention and success (Kramer-Koehler et al. 1995; Fromm 2003). However, there has been little detailed and specific guidance about gender inclusivity available for educators in engineering and related sciences. Hence the need for this book.

It is neither possible nor sufficient to change the curriculum in engineering by imposing a top-down ruling from management that this will happen. Change will not occur until those responsible for curriculum development and teaching are convinced that there is a need for change which will benefit all students and faculty members and the profession as a whole. Overcoming the resistance of faculty members to make changes will be discussed in more detail in Chapter 7, but there may also be resistance to change from students themselves, predominantly male but also from some of the female students. As part of the masculine culture that currently exists, there can be 'a certain competitive arrogance and even contempt for non-technical material (often unhelpfully described as 'soft-skills') and of students majoring in such disciplines . . . Such attitudes certainly inhibit change . . . and may continue to undermine curriculum and other innovations' (King 2008: 72). Overcoming these attitudes and resistance to change will be examined in subsequent chapters.

ACCREDITATION CRITERIA AND CURRICULUM FRAMEWORKS

In all countries that provide engineering education at tertiary level, the profession provides some form of accreditation requirements for engineering programs. The accreditation may be carried out by the professional engineering organizations or by an independently incorporated body and will generally occur on a designated time cycle, such as every five years. The accreditation process will usually involve some form of documentation being developed by the education institution to demonstrate the way in which their engineering programs meet the accreditation criteria. In addition the process also usually involves a visit to the campus by representatives from the accrediting body to view the physical facilities and to conduct extensive interviews with managers, faculty members and students of the engineering programs. In an increasingly globalized profession, accreditation is one of the means by which qualifications gained in one country can be recognized in another, enabling engineers to work all over the world. To this end the International Engineering Alliance (IEA) was formed in 2007 and incorporates international mobility agreements for engineering qualifications and practising engineers at all levels of the profession.

Accreditation requirements are sometimes used as an excuse for not changing the engineering curriculum to become more inclusive. In fact this is almost certainly not the case, as the majority of accreditation criteria have become increasingly outcomes focused rather than prescriptive and have therefore opened up opportunities for significant curriculum innovation. This book is about promoting change in an ongoing way, not only in terms of individuals or courses but rather systemic change ensured by accreditation procedures.

CHAPTER OVERVIEW

In Chapter 2 we examine the current and historical situation for women in engineering at university and within the profession, based on current statistics and previous research. Chapter 3 reviews the range of non-curriculum based initiatives that have been employed by higher education institutions, governments and professional organizations to encourage women to commence engineering degrees and to complete those studies.

Chapter 4 discusses in more detail the theories of learning, understandings of curriculum and the theory of curriculum applied to gender inclusive engineering education. It demonstrates the importance of considerations of identity, self efficacy and learning styles in achieving effective learning and teaching outcomes. Evidence of the effectiveness of gender inclusive curricula in engineering is also examined.

The focus of Chapter 5 is on the practical design and implementation of gender inclusive curriculum in engineering. We go beyond the use of gender inclusive teaching methods and also discuss inclusive design of the curriculum. This chapter is aimed at faculty members who teach engineering, either as individuals or in teams. It incorporates many practical suggestions and examples. Chapter 6 provides many further examples of gender inclusive curriculum in a wide range of undergraduate engineering courses covering all year levels across many engineering disciplines.

Chapter 7 focuses on the ‘big picture’ issues of implementing gender inclusive engineering education at a university, department and program level, including the need to provide training and development for faculty members. It offers advice and strategies for embedding these principles into course and program approval processes at universities and accreditation requirements of professional bodies, to ensure that their implementation is sustainable. The evaluation of success is also discussed.

Chapter 8 provides a summary of the whole argument and suggests some ways of reframing engineering education that can be readily implemented. The approach taken in the book allows application across international boundaries, with a particular focus on the US, Canada, UK, continental Europe and Australia.

2 Women in Engineering
The Current Position

INTRODUCTION

In this chapter we provide some statistics and accompanying discussion to describe the current position of women in engineering in both the higher education and professional workforce sectors. We consider the situation in a range of different countries and regions of the world, as well as over a range of different engineering specializations. There is some focus on the English-speaking countries, particularly Australia, Canada, UK and US, because, as will be seen, these countries have lower participation rates of women in engineering in nearly every statistic examined. Although we present workforce statistics as well, and briefly discuss some of the issues and programs used to try to retain women in the professional workforce, a more detailed discussion of this area is beyond the scope of this book, and references have been provided for readers who are interested in pursuing this topic further. In subsequent chapters we explore in greater detail the issues that women in engineering education face and the programs that have been used to attract and retain women in engineering.

WOMEN IN ENGINEERING EDUCATION AT TERTIARY LEVEL

Tertiary level is used in this book to describe post–compulsory education that is undertaken either at a university or technical college of some type, after the completion of high school (see Appendix B). Engineering education is provided at a range of levels which result in different qualification names that correspond with a range of different occupational levels within engineering. The qualification names vary between countries. For example in the Australian and other British education systems, a diploma in engineering is a vocational level qualification provided at a technical college, usually of two to three years' duration. It is intended to provide entry to engineering at a para-professional level in roles such as draftpersons or technical officers. A diploma is not recognized as sufficient for professional registration as a graduate engineer. However, in many European countries (France and Germany for example) a diploma is the name of the usual qualification required for entry to full professional engineering status, requiring five years of study at a prestigious university. In many other countries, such as the US, Canada, Australia and Japan, the equivalent minimum qualification required for professional registration as a graduate engineer is a bachelor's degree of four years' duration.

Recently some attempts have been made to standardize these qualifications across a number of European countries. The Bologna Process that commenced in 1999 has the aim of ensuring consistent approaches to nomenclature, duration and level of higher education qualifications across 46 European countries. The overarching framework of qualifications of the European Higher Education Area (EHEA) was adopted by the Ministers of Education of the Bologna Process at their meeting in Bergen in May 2005 and further endorsed at the London meeting in 2007. The Bologna Process agreement defined the following levels of professional engineering undergraduate and postgraduate qualifications:

- 1st cycle: typically 180–240 ECTS (European Credit Transfer and Accumulation System) credits, usually awarding a bachelor's degree.
- 2nd cycle: typically 90–120 ECTS credits (a minimum of 60 on 2nd cycle level). Usually awarding a master's degree.
- 3rd cycle: Doctoral degree. No ECTS range given.

A typical full year of study involves completion of 60 ECTS credits of study; hence in most cases these qualifications will take three to four, one-and-a-half to two, and three years respectively to complete. The actual naming of the award may vary from country to country. By 2010, all member countries are required to develop national qualifications frameworks that are compatible with this overarching framework. This requirement has led to progressive changes in engineering degree offerings in many countries (EHEA 2005).

The current range of qualifications and standards causes some difficulties in comparing statistics from different countries on what constitutes tertiary level engineering education, particularly when looking at historical data. For this reason the statistics presented here focus on full professional engineering qualifications which require a minimum of at least three but usually four years of study at a university (refer to Appendix B for university definition). For ease of presentation in this book, this qualification is referred to as undergraduate, bachelor level study. Post-graduate statistics at doctoral level are also included, as these are generally considered to be the first research level qualification. Master's level statistics are excluded due to the wide variety of definitions of this level of study in different countries.

A further difficulty in comparing statistics from different countries is created by the different definitions used by various bodies to define engineering study. In many instances the field of education may include additional areas of study, and these, combined with engineering, can slightly blur the picture. For example the Organisation for Economic Cooperation and Development (OECD) statistical database (OECD 2008) uses *Engineering, manufacturing and construction* as a category, which is defined by the International Standard Classifications of Education (ISCED 1997) to include three subcategories of *Engineering and Engineering trades; Manufacturing and processing;* and *Architecture and building*. The standard disciplines of engineering are scattered across these three subcategories: electrical and chemical engineering are in the first, industrial engineering in the second and civil engineering in the third. Thus, none of these subcategories alone gives a true picture of engineering as a whole. In addition, the third category includes the professions of architecture and town planning which attract significantly more women than does engineering and hence will give an over-estimate of the female participation percentage in this category. Consequently, neither the overall *Engineering, manufacturing and construction* category nor any one of the subcategories gives an accurate picture of professional engineering numbers and will tend to over-estimate these figures. The National Science Board (NSB) in the US incorporates the OECD statistics and others from different sources under the single label of *Engineering*, which is then not clearly defined (NSB 2008), so this is likely to over-estimate percentages even more. A further problem is that statistics are available for different years in different countries, making same-year comparisons difficult. Statistics are presented for the latest year available in each case, but some historical discussion is also included. Where possible, more accurately defined statistics for professional engineering graduates and employment are used, particularly those available from engineering education or professional engineering associations in the relevant countries.

The statistics presented also specifically exclude vocational or para-professional level qualifications in engineering. This is not to say that the lack of women is less pronounced or that their presence is not equally important at that level, but rather that the range of statistics published in those areas is more limited and that the range of definitions is much more varied between countries than even bachelor level qualifications, making meaningful comparisons more difficult.

Bachelor Qualifications in Engineering

Completion or Graduation Statistics

Statistics for undergraduate education may include access (or commencing) students, total students and graduating (or completing) students. A comparison between percentages of students commencing and completing engineering study can be used to indicate the retention of students, namely the proportion of those who originally enrol who go on to complete their engineering degree within the required time frame. However, graduation statistics are the most readily available and these are summarized here, with some additional discussion of retention statistics where they are available.

Tables 2.1 and 2.2 summarize the most recent statistics available in a range of countries broadly representative of their regions (Table 2.1) and regional averages as a whole (Table 2.2) for the percentage of women in the total numbers of students completing engineering study at bachelor level. Wherever possible these statistics are drawn from direct sources within the relevant country; for example US statistics are from the American Society of Engineering Education, an organization that publishes annual statistical summaries on numerous aspects of engineering students and faculty members. Where in-country statistics are not available (or not readily accessible in English), three major sources of data are used, namely the OECD *Education at a Glance,* 2008 and the associated *Online Education Database* (OECD 2008); the NSB *Science and Engineering Indicators 2008* (NSB 2008) and the National Science Federation (NSF) *Women, Minorities and Persons with Disabilities in Science and Engineering: 2009* (NSF 2009). However, it should be noted that where in-country statistics were available, they consistently provide percentages of women that are lower than any of the statistics from these three sources, almost certainly due to the slightly broader definitions for engineering categories in the first two publications as discussed previously. Hence it should be considered that those statistics from the OECD, NSB and NSF are over-estimates.

Examining historical data trends also helps to put these statistics into perspective. The US percentage of 18.1 per cent total undergraduate completions that were female in 2007–2008 was the lowest percentage since 1996 (ASEE 2009). In Canada, female graduation from undergraduate engineering degrees decreased from a peak of 21.8 per cent in 2003 to 18.9 per cent in 2007 (Engineers Canada 2009). In Australia the comparative figures were a peak of 17.4 per cent in 2003, down to 16.2 per cent in 2007, below the levels of 2001 (DEST 2009). The percentage of female graduations in engineering in the UK has not followed this trend, steadily increasing to 15.5 per cent in 2005–2006 (UKRC 2009c), but this is still noticeably lower than in comparable countries. Hence the overall picture in the English-speaking western world in recent years is one of declining percentages of women completing engineering study, even though peak percentages were never high. This may partially be explained by female numbers of graduates holding steady whilst male graduate numbers are increasing, but why is this occurring?

As indicated in Tables 2.1 and 2.2 there are some countries and regions where the percentage of women graduating with engineering degrees is significantly higher than in the US or UK, although nowhere is it in line with the female percentage of overall population, or of the female percentage of university graduates. The formerly communist countries of Central and

Table 2.1 Percentage of All Students Graduating with a Bachelor Degree in Engineering Who Are Women—Representative Countries

Region and Country	% of Women	Year of Data	Data Source
Americas			
United States	18.1	2007–2008	ASEE (2009)
Canada	18.9	2007	Engineers Canada (2009)
Mexico	27.2	2006	OECD (2008)*
Brazil	31.4	2005	OECD (2008)
Europe			
United Kingdom	15.5	2006–2007	UKRC (2009c)
France	25.6	2005	NSF (2009)*
Germany	22.2	2005	NSF (2009)
Denmark	29.5	2005	NSF (2009)
Poland	28.4	2005	NSF (2009)
Slovak Republic	31.8	2005	NSF (2009)
Middle East & Africa			
Turkey	22.0	2005	NSF (2009)
Israel	26.0	2006	OECD (2008)*
Ethiopia	6.5	2004	NSB (2008)*
Asia & Oceania			
Japan	11.0	2006	OECD (2008)
Australia	16.2	2007	DEST (2009)

*See earlier discussion about over-estimates in OECD, NSB and NSF statistics due to definitions.

Eastern Europe are generally considered to have the highest proportion of female engineers in the world, and this has long been the case. For example in 1967 women accounted for 34 per cent of engineers in the Soviet Union, while at the same time the percentage in most western countries

Table 2.2 Percentage of All Students Graduating with a Bachelor Degree in Engineering Who Are Women—Regional Averages

Region	% of Women	Year of Data	Data Source
Asia	15.9	2004	NSB (2008)*
Middle East	17.6	2004	NSB (2008)
Africa	14.2	2004	NSB (2008)*
Europe	24.1	2004	NSB (2008)
Central/Eastern Europe	30.4	2004	NSB (2008)
Central/North America	22.1	2004	NSB (2008)*
South America	29.4	2004	NSB (2008)
Oceania	25.3	2004	NSB (2008)

*See earlier discussion about over-estimates in NSB statistics due to definitions.

was negligible (Allen and Robins 1990). However, it should also be noted that the status of engineers in the Soviet Union was not the same as elsewhere, with one author stating that 'Soviet engineers, especially women, have long been little more than workers' lackeys' (Malysheva 1992: 197). In any case, since the political and government changes after the disbanding of the Soviet Union, there has been a dramatic decrease in the percentage of women engineering students. In Russia this has dropped from 58 per cent in 1995 to 30 per cent in 2002 (Davidson and Burke 2004). It is difficult to obtain reliable statistics from China, but a 1994 report stated that women accounted for 27 per cent of engineering college graduates in 1992 (Information Office 1994). As China increases its economic and cultural ties with the western world, it seems unlikely that this percentage will increase. There are also some smaller countries with higher percentages of female engineering graduates, such as Iceland, but the overall numbers from these countries are very small and thus the female percentage is subject to high variation. It is likely therefore that the percentage of female engineers graduating across all countries will remain stationary in the range of 15 per cent to 30 per cent at best for the near future.

Distribution Across Engineering Disciplines

There are some engineering specializations that have higher percentages of female students than the average percentage across all engineering specializations. Examples often cited are chemical engineering, environmental engineering, biomedical engineering and industrial engineering (which is the US term for an area often called production or manufacturing engineering elsewhere, that may not be recognized as a separate discipline in other countries). Table 2.3 presents some data on this issue, although data are not widely available for all countries.

It should also be remembered that those specializations with higher percentages of females frequently represent a very small percentage of the overall number of engineering degrees awarded, since the traditional disciplines of civil, electrical and mechanical engineering still make up the vast majority of degrees awarded. For example chemical and environmental engineering degrees, with over 40 per cent women, accounted for a total of only 9.3 per cent of all engineering degrees awarded in Canada in 2007. At the same time electrical and mechanical degrees, with low percentages of women, accounted for 70.4 per cent of all engineering degrees (Engineers Canada 2009). Hence, whilst there may be lessons to be learned in terms of the content and curriculum of the degrees which are attractive to women, it will not make much difference to the overall numbers of women engineers unless the larger, more traditional discipline areas can attract more women.

Retention, Success and Persistence in Engineering Study

Retention rates are defined in a range of ways by different countries and organizations. A fairly standard but conservative definition is the percentage of all students who enter an engineering degree program who graduate from it within the standard time frame. This is also sometimes defined as the success or persistence rate. However, this definition usually does not capture those students who graduate but take an additional year or two

Table 2.3 Percentage of All Students Graduating with a Bachelor Degree in Engineering Who Are Women—by Engineering Discipline, Selected Countries

Discipline	USA	UK	Canada
Civil	21.1	15.8	22.3
Chemical	34.9	28.8	42.3
Electrical	12.1	11.6	15.3
Environmental	43.2	n.a.	46.4
Industrial/production	30.4	16.9	23.6
Mechanical	11.9	7.6	12.7

Statistics from Gibbons (2008), UKRC (2009c), Engineers Canada (2009).

to do so, which is a relatively common situation in engineering programs where students may struggle to complete early mathematics and science courses that are sometimes used as gatekeepers for the degree. This retention statistic definition is also unable to capture students who transfer into an engineering program part way through, either from another institution or another degree program for which they receive some credit.

Engineering programs have long been noted as having relatively poor retention or success rates, when compared with other university programs. Reported rates in the US range from 40 to 70 per cent depending on exact definition and institution (Hartman and Hartman 2006). Adelman's 1998 US study indicated that the retention rate for female engineering students was consistently lower than that for males, even when their results were similar to or better than those of the males. However, a more recent study based on the examination of nearly 70,000 engineering students from nine institutions in the US, found that 57 per cent of students persisted in engineering to the end of their fourth year (Ohland et al. 2008). This was higher than all other major groups considered, including business (55 per cent) and computer science (38 per cent). Ohland et al.'s study also found that the persistence rate of women in engineering was similar to that of men, which differs from the earlier studies.

Australian data indicate that the four-year graduation success rates for engineering students in 2005 were 51.3 per cent for males and 57.8 per cent for females (King 2008). This result would seem to indicate that female students, although a minority, are more likely to complete the course than are men. Retention statistics for other countries are hard to find. The UK publishes non-continuation performance indicators in higher education, but these are not gender disaggregated. Data are gathered for a single year and then projected for graduation. Using this measure the projected percentage of students enrolled in engineering in 2006–2007 across all UK universities who will obtain their degree is 77 per cent (HESA 2009). A related report considering one year continuation rates shows that engineering continuation rates are 88.5 per cent compared to the overall rate of 90.6 per cent (National Audit Office 2007). Hence we could extrapolate to predict likely success rates of 75 per cent for engineering graduation.

Why is consideration of retention rates so important? One reason is that there have been a considerable number of efforts over many years focussed on attracting students, both male and female, to study engineering. These efforts have often focussed on the encouragement of upper year high school students to undertake the mathematics and science study that is required for entry to engineering degrees. However, a significant amount of this effort is wasted if between 25 to nearly 50 per cent of students who have jumped the entrance hurdle, often with higher grades than university entrants in other fields (Seymour 2002), do not then complete their engineering degrees. The reality is that the overall retention rates in university engineering study are poor in all countries, even if they are only as bad or slightly worse or better than average retention rates for all tertiary study.

One of the clear trends within retention statistics is the very wide variation between rates at different institutions in the same country. For example, even though Ohland et al.'s (2008) study encompassed a large number of students (N = 70,000), the persistence rates in engineering at the nine institutions in the main dataset actually varied between 37 per cent and 66 per cent, with higher retention rates at the institutions with larger student numbers leading to the overall average rate of 57 per cent. The significant variation between institutions warrants further investigation of the ways in which the engineering education curriculum might operate to promote stronger student retention and completion.

Similarly the UK projected continuation rates vary from 52 per cent to 97 per cent. Some of these variations are explained by the particular circumstances of the student cohort. For example, universities with higher percentages of mature-age students or racial minority students often have lower retention rates and this may be for a range of complex economic and social reasons. It is important therefore, in order to improve our understanding of women's persistence and success in engineering study, to examine the practices of those universities which maintain higher retention rates, particularly where they have student cohorts with a higher than average proportion that may be considered disadvantaged by social, economic or educational background.

As stated earlier the data on retention rates of women in engineering are somewhat contradictory. The most recent data in the US (Ohland et al. 2008) and Australia (King 2008) indicate that female retention rates are the same or higher than male retention rates. One possible explanation is that, in spite of the known difficulties that they face in the male dominated environment, the small number of women who enter engineering are a self-selected, highly resilient group. This feature was one of the findings of a US study that also noted that the women who continue in, and graduate from, engineering "are likely to have strong family support, high expectations, healthy self-confidence, and solid academic preparation" (NCES 2000: 88).

For some time it has been clear that one of the key reasons that all students, whether male or female, leave engineering is that they become disengaged from their study, and that dissatisfaction with teaching, curricular factors and the traditional pedagogy of engineering programs are the major reasons for this (Seymour and Hewitt 1997; Adelman 1998). Capability is not the issue, as those students who leave engineering are shown to be as academically able as those who continue (Seymour and Hewitt 1997). Disturbingly, recent findings indicate that even though engineering students actually persisted in their study at a higher rate than students in other fields, 'their academic disengagement increases over the course of their undergraduate education' (Ohland et al. 2008: 273). Clearly this situation is another indication of the significant room for improvement in the engineering curriculum.

Doctoral Qualifications in Engineering

Doctoral degrees in some fields of science, such as chemistry, physics and biology are considered almost essential in many countries for graduates to be able to enter one of the significant employment options in these fields, namely that of research and development. Consequently the percentage of females enrolling and completing doctoral degrees in these and similar fields is critically important from an equity perspective. Nevertheless, in engineering a doctoral degree is not required to enter professional practice, and the proportion of total engineering employment in the research and development sector is quite small. Hence obtaining a doctoral degree in engineering is primarily relevant for those graduates who wish to pursue a career in academia or research. The presence of female faculty members as role models can be an important factor in retaining female undergraduate students (NSF 2005), so the percentage of women enrolling in and completing doctoral studies in engineering is another important part of the overall picture of women in engineering that needs to be considered. Table 2.4 presents the data available in this category for a range of countries.

Apart from careers in academia, doctoral graduates may also be employed in research. The European Commission has noted that gender imbalance is not the only issue of concern for females in engineering and scientific research. They assert that there is clear evidence that women in these fields are under-deployed, lower paid, have poorer access to resources and significantly less chance than men of reaching senior or influential levels. They also argue that it will be difficult to sustain research capacity if intellectual resources are not sourced and supported on a more equitable basis (European Commission 2006).

Women Faculty

The presence of women faculty members in engineering schools is an important way to provide role models and mentors for female undergraduate and graduate students (NSF 2005). Conversely, the lack of female peers and role models can be discouraging for female students (Goodman Research Group 2002). However, the number and percentage of women faculty members in engineering is very low overall, particularly at the higher levels of full professor. In the US in 2008 women made up only 12.3 per cent of total tenure track faculty in engineering and only 7.4 per cent of full professors (Gibbons 2008). Similarly in Europe, only 5.8 per cent of full professors in engineering were female (European Commission 2006). In addition, US studies indicate that women faculty earn less, are promoted less frequently to senior academic ranks and publish less frequently than their male counterparts (NSF 2005).

Table 2.4 Percentage of All Students Graduating with a Doctoral Degree in Engineering Who Are Women—Selected Countries and Regions

Region and Country	% of Women	Year of Data	Data Source
Americas	17.6	2004	NSB (2008)
United States	20.8	2007–2008	Gibbons (2008)
Canada	15.8	2007	Engineers Canada (2009)
Mexico	20.3	2004	NSB (2008)
Europe	17.1	2004	European Comm (2006)
United Kingdom	19.9	2006–2007	UKRC (2009c)
France	22.8	2003	European Comm (2006)
Germany	6.8	2003	European Comm (2006)
Denmark	23.8	2003	European Comm (2006)
Romania	37.5	2003	European Comm (2006)
Middle East	21.8	2004	NSB (2008)
Turkey	14.7	2003	European Comm (2006)
Israel	31.0	2004	NSB (2008)
Asia & Oceania			
Japan	10.1	2004	NSB (2008)
Australia	20.3	2007	DEST (2009)

WOMEN IN THE PROFESSIONAL ENGINEERING WORKFORCE

In many countries the percentage of women undertaking engineering study at tertiary level was almost zero until the 1970s, still very low in the 1980s and only gradually increased to current levels from the mid to late 1990s. Consequently the percentage of women in the professional engineering workforce is even lower than that in engineering study, since many males who graduated in the 1960s and 1970s are still participating in the workforce. These low percentages also mean that women may be represented in very small numbers in individual workplaces, with the situation of being the lone woman engineer at a company or work site still not unusual.

The same issues pertaining to the difficulty in comparing statistics from different countries that applied in the education area also apply to statistics in the professional workforce, namely different categories and groupings for engineering occupations, different definitions of professional engineers, different years of data collection and so on. For these reasons we should be wary of comparing statistics too closely, but the overall trends tend to mirror those noted in the education statistics.

Professional Engineering Workforce Participation

Table 2.5 illustrates the latest data available for the percentage of women in the professional engineering workforce in the same countries considered in the education statistics, where available. Similar to the education statistics, the definitions are variable in the different datasets and in some cases over-estimate the engineering percentages due to the inclusion of other professions in the occupation category. For example the European statistics from UKRC (2009b) use the classification *Engineering professionals* based on the International Standard Classification of Occupations, but this includes *Architects, engineers and related professionals*, and hence will over-estimate the female percentage.

Women's Progress, Success and Retention in the Professional Engineering Workforce

While the professional workforce participation percentages of women in engineering are small in almost all countries, retention statistics indicate that maintaining even these small numbers of women in the engineering profession is an ongoing problem. A major report on women in Science, Engineering and Technology (SET) commissioned by the US Congress indicated that women are twice as likely as men to leave engineering and science careers and also have much higher exit rates than women in other professions (CAWMSET 2000). A more recent US study (Hewlett et al. 2008) found that 52 per cent of women scientists, engineers and technologists leave their professions, particularly when they are in their mid to late 30s. Data collected in 1996 by the Association of Professional Engineers, Scientists and Managers, Australia (APESMA) found that 18 per cent of all engineering graduates were female. However, by 2006, female engineers made up only 11 per cent of all engineers with between seven and 10 years' experience, which indicates that women are leaving the engineering profession significantly more quickly than their male counterparts (APESMA 2007).

Why are so many women leaving the engineering profession after they have persevered to obtain their engineering degrees and gain entry to the profession? Numerous studies show that women in the engineering workforce may face a range of problems that disadvantage them when compared with their male colleagues. The sum of these accumulated disadvantages

Table 2.5 Percentage of the Professional Engineering Workforce Who Are Women

Region and Country	% of Women	Year of Data	Data Source
Americas			
United States	11.1	2007	NSF (2009)
Canada	12.2	2006	CCWESTT (2008)
Europe			
United Kingdom	8.7	2007	UKRC (2009b)
France	16.8	2007	UKRC (2009b)
Germany	15.4	2007	UKRC (2009b)
Denmark	15.4	2007	UKRC (2009b)
Romania	25.3	2007	UKRC (2009b)
Middle East			
Turkey	27.0	2008	Denzig (2008)
Asia & Oceania			
Japan	6.2	2004	Burrowes (2006)
Australia	11.2	2007	ABS (2009)

eventually leads to significant numbers of women leaving the profession for other employment or leaving the workforce altogether.

In many cases these disadvantages in employment prospects and conditions commence as soon as these women graduate. US statistics indicate slightly lower initial employment rates for women in engineering with 5.8 per cent of women engineering graduates from 2003 to 2005 unemployed in 2006, compared with 3.2 per cent of men. Salary data for full-time employed engineers aged 29 and younger in 2006 indicate median salaries of $US56,000 for men and $US55,000 for women (NSF 2009). This initial pay gap worsens over time, with women engineers in the 40 to 49 year age group having a median salary of 93 per cent of the salary of male peers. Worse pay gaps exist in the UK with female engineering professionals in 2008 earning a gross salary of £30,290, which was only 81 per cent of the male gross salary of £37,559 (UKRC 2009a). German data show that female engineers and computer scientists start with lower level positions and salaries, have double the unemployment rate of females in other professions and earn only 70 per cent of their male colleagues' salaries after five years. They also have lower rates of promotion and lower rates of acceptance in the profession (Femtec n.d.).

In response to the poor success and retention outcomes for women in engineering, companies and governments have attempted to implement numerous programs and measures to stem the tide, although often with limited success. A full discussion of these issues and programs is beyond the scope of this book, but readers seeking further information about such programs and research in this area are referred to the following: Greenfield et al. 2002; Womeng 2005; Mattis 2007 and Hewlett et al. 2008.

SUMMARY

We have presented comparative data on women's representation in engineering education and employment in countries across the world. There are a number of difficulties in collating reliable and consistent data which need to be borne in mind when making trans-national comparisons, such as the different terminologies and categorizations of 'engineering', lack of parity in qualifications required for professional engineering status, and the different years over which the latest data from different countries are available. Nevertheless, the overall picture is that women are still severely under-represented in engineering. In the principal English-speaking countries, female engineering graduates range from 15 per cent to 19 per cent of total engineering graduates and this statistic has been falling in recent years. In contrast, in the Eastern bloc countries, the comparable statistic approaches 30 per cent, but in some countries this is less than it previously was. There is wide variation in these data across engineering disciplines. In chemical, environmental and biomedical engineering for example, proportions of female engineering graduates in the US, UK and Canada range from 35 per cent to 46 per cent approximately. However, total numbers in these degree programs are very small in comparison with the three main engineering disciplines: civil, electrical and mechanical, where the comparable data range from 7.6 per cent to 22.3 per cent and are strongly skewed to the lower part of this range. Consequently, it is principally these three engineering fields which will have to attract and retain more women students if there is to be any significant improvement in female representation in engineering as a whole.

Comparative gender-disaggregated data on retention of engineering undergraduates are even more difficult to obtain than information about proportions of women in the student body. Until very recently, the available data indicated that female attrition rates were higher than those for males. A recent large US study (Ohland et al. 2008) appears to contradict this, finding no difference between male and female retention rates: about 57 per cent for both. However, their findings with respect to gender are noted as preliminary, and the authors are currently investigating gender issues in their dataset more closely. One clear factor in retention data is the wide variation that exists between institutions. For example, in the nine institutions included in the US study, the retention rates ranged from 37 per cent to 66 per cent, a clear indication that some institutions are more successful than others in retaining engineering students.

Looking beyond the undergraduate stage, the percentage of women obtaining doctoral degrees is similar to those obtaining undergraduate degrees, though with some interesting variation between countries. In the engineering workforce, in the English-speaking countries and Western Europe, the participation rates of women range from 11 per cent to 17 per cent, and in the Eastern bloc they average 25 per cent. Worryingly, it has been found that women are far more likely to leave the profession than men and that this is due to a range of disadvantages that they face, not least of which is a significant pay gap.

In summary, women remain under-represented in engineering, at undergraduate and postgraduate level and in the workforce. Some of the most severe under-representation is in the English-speaking countries. Furthermore, undergraduate enrolments in these countries are declining, and additionally, much research has shown that women students are more likely to leave their engineering studies than men. Many studies show that the main reason male and female students leave engineering is that they become disengaged and demotivated about their studies. There is a clear message here that the undergraduate engineering curriculum is failing to inspire our female and male students. The remaining chapters of this book examine the reasons for this and suggest solutions.

3 Attracting Women to Enter and Succeed in Engineering Study

INTRODUCTION

The statistics presented in the previous chapter show that women remain severely under-represented in the engineering professional workforce and constitute approximately one-quarter or less of engineering students in the majority of countries and regions around the world. In light of this, many people, institutions, professional organizations and government bodies have developed, implemented and supported a multitude of measures and programs designed to increase the attractiveness of engineering study to women and then to support them while they undertake it. This book presents a different approach. Its main focus is on the need to develop a gender inclusive curriculum in engineering and the provision of practical advice on how to do that, in order to achieve the goals of increased participation and success of women in engineering education. However, we also acknowledge that whilst curriculum change is a critical (and up to now neglected) component of gender inclusive engineering education, the measures described in this chapter are also critically important if we are to make progress towards the goal of increased participation and success for women in engineering.

We begin with an examination of the theoretical contexts of many of the past and present approaches to increasing the numbers and improving the outcomes for women in engineering education. We then look at a wide range of programs and initiatives in the following sections, structured around the terms 'access', 'retention' and 'success'. These terms were also used in Chapter 2, where access was defined as the number of students commencing engineering study. In this chapter we use a general definition of retention as maintaining the enrolment of students in engineering programs until they graduate with an engineering degree and a definition of success as the level of an individual's achievement within that degree in terms of their final grade (often expressed as a grade point average). We consider programs that are developed by schools, universities and professional organizations that are aimed at targeting individual women, women as a group within the student body, teachers in schools and faculty members in universities. We include many individual examples as well as some specific studies of institutional or organizational initiatives in various countries to attract, retain and support women in engineering education. We conclude the chapter by considering the evaluation of these programs and initiatives in terms of their success and aspects that have been left unaddressed.

THEORETICAL CONTEXT

The representation and influence of women engineers depends, firstly, on how many enter tertiary education to study engineering; secondly, on how many successfully complete their studies and subsequently how many enter, stay in, and rise to seniority in the engineering profession. Students' choices about tertiary study, and university policies and practices which support or impede the admission of particular groups of students, are access issues. The retention and success of female students are influenced by how their learning experiences affect their persistence with their studies in engineering, along with the usual range of factors that impact on student retention.

Historical Development

Early research on women's access and success in science, engineering and technology (SET) tended to focus on the barriers faced by these students in entering and participating in tertiary education (Cross 1984; Whyte 1986; Kelly 1987). Cross (1984) summarized these barriers as situational, institutional and dispositional, where situational barriers are those which arise from an individual's life circumstances at a particular time, institutional barriers are the formal policies and procedures which impede the entry or success of minority student groups such as women in engineering and dispositional barriers are those arising from students' own attitudes and perceptions which affect their choice of field of study or their success within it. Among the barriers cited with regard to engineering are sex bias within vocational guidance; a lack of knowledge among teachers and parents about women's ability to succeed in modern engineering; failure of professional associations to take responsibility for the problem; the 'masculine' image of engineering and the perception that it is incompatible with 'normal femininity' and family roles; lack of female role models and male hostility and perceptions that women are incapable of, or unsuited to, engineering (Bureau of Labour Market Research 1985; Maskell-Pretz and Hopkins 1997; see overview in Bastalich et al. 2007). Several studies have highlighted the failure of engineering to attract women at secondary school level because of its comparative invisibility and negative experiences of physical science at school (Bureau of Labour Market Research 1985; Jawitz et al. 2000; Chiu et al. 2002). Some authors also comment on the need to improve women's confidence and success in preparing for engineering courses or engineering professional practice (Ayre and Beynon 1988; Maskell-Pretz and Hopkins 1997) and the influence of women's childhood socialization on their acquisition of technical skills (Hacker 1989; McIlwee and Robinson 1992). These findings formed the basis of many strategies of the period designed to increase the participation of girls and women in mathematics and related disciplines by demonstrating the usefulness and relevance of science and engineering, and challenging these disciplines to be more inclusive (Harding 1996). However, many of the authors of these strategies assumed that once women had been assisted to overcome these barriers, they then would be assimilated into the engineering culture and no further action would be required.

The studies that focussed on barriers for, or differences of, women have been criticized since the 1990s (for example WISET 1995; FEFC 1996; Cronin et al. 1997; Gill, Sharp et al. 2008) on the grounds that they emphasized ' . . . deficiencies in ability, preparation, or socialization of women which rendered them unlikely to take up education and careers in a system organized and dominated historically by men' (Gill, Sharp et al. 2008: 393; and see Sonnert 1999 for a full discussion). In these approaches, engineering is maintained as value neutral with the implication that, once barriers and discrimination preventing or obstructing women's participation have been removed, they will be free to compete on equal terms with men. The outcome in practice tends to be an emphasis upon the introduction and remediation of women without changing the way engineering is taught and practiced (Blickenstaff 2005; Rosser 1998). However, Bastalich et al. (2007) note that a study of US faculty programs by Fox (1998) found that, even when programs for change are based on the assumption that the problem is grounded in the engineering and science culture, solutions tend to target women and the creation of alternative, supportive spaces for them because this is an easier and less controversial strategy than attempting institutional change (Fox 1998). Despite these shortcomings, both the proportions of women students studying engineering and the absolute numbers rose in the 1970s and 1980s, and we therefore believe that these studies were a valuable contribution to the field and reflected the time in which they were undertaken, when traditional approaches to equal employment opportunity (EEO) policies targeted women as the source of the problem (Bacchi 1999).

Instead of the 'blaming the victim' approach which focusses on a female deficit, other researchers (Harding 1991; Bianchini et al. 2000) looked instead for socio-cultural explanations for the lack of women in engineering and science classes, for example ' . . . the intersection of professional and personal identities; views of the nature of science; beliefs related to students' experiences in science education; and kinds of curricular and instructional strategies . . . ' (Bianchini et al. 2000). Using this approach we suggest that the problem can be understood as an emergent property of the situation that begins in school—girls as a minority in senior mathematics and science classes, women as a minority in the engineering student body and in engineering workplaces. Thus we should focus on social systemic issues such as the gendering of certain school subjects, the lack of senior women role models together with the (taken for granted) assumptions about men's work and women's work, which combine to orient girls and women away from careers in engineering. This approach adopts the position that individual and environment are inevitably interrelated and thus avoids the heavily individualized focus of some of the previous work.

Access of Women to Engineering Education

McIlwee and Robinson (1992) researched the problems and opportunities for women in engineering in the US in the late 1980s by investigating the life cycles of some women engineers from high school through to career issues such as balancing family and professional life. The three factors identified by these authors as being most influential in steering women into SET were, in order, academic abilities, pragmatic career choice and personal relationships. In contrast, the correspondingly powerful influences for men were tinkering experience, encouragement by family members, followed by ability in mathematics and science. Other issues that must be considered are the entry requirements to SET higher education courses, and whether there is institutional support for alternative entry routes for students whose previous educational experiences have not included science subjects at senior secondary level.

Jolly et al. (2004) suggest that access issues can be classified into three essential groups of prerequisites enabling students to participate successfully in quantitative fields of study which are not necessarily equally weighted for any individual. These categories are:

- engagement: interest, inclination and motivation to study in this field
- capacity: the necessary knowledge and skills
- continuity: access to the necessary institutional programs and resources.

These categories are used for the following discussion of the literature on access issues.

Engagement: Interest, Inclination and Motivation

The importance of family and friends on women's choice of SET degree study has been identified by a number of authors (Carter and Kirkup 1990; McIlwee and Robinson 1992; Geppert 1995; Hobart et al. 2006). McIlwee and Robinson found that the influence on both men and women to become engineers was particularly strong if their fathers were engineers. Thirty-five per cent of the women engineers in their sample had fathers who were engineers, compared with 27 per cent of the men. In a more recent Australian study involving interviews with 41 women and 10 men engineers across a range of discipline and geographical areas in Australia, and with a range of years of professional engineering experience (Mills et al. 2006; Gill, Sharp et al. 2008), 20 out of the 41 women interviewed had a family connection with engineering; only one stated that her father was opposed to her choice and nine of the women stated that their father was an engineer.

> I always liked science and I'd always done science but at that stage if girls did science you became a science teacher and I didn't want to teach. My dad worked for the airline as a maintenance engineer and I think that probably influenced me.
>
> (Gill, Sharp et al. 2008: 396)

Sometimes it was another family member or friend who had affirmed the potential pleasure of getting involved in engineering work. Women with family connections were able to sustain the idea that their talents were a natural or even genetic product, and they also had a better idea of what an engineer does than those who had no connection with the field. Another obvious advantage was that they came into a culture about which they had some prior knowledge, and they were more prepared for the work environment than some of their female peers (Gill, Mills et al. 2008).

Several authors have also noted that teachers, careers advisors and friends have had a greater influence on women than on their male counterparts (McIlwee and Robinson 1992; Evetts 1996; Capobianco 2006). Teachers had often been inspiring motivators who had led them into taking up engineering at university. In the Australian study (Gill, Mills et al. 2008) 11 of 41 women, but only one of 10 men, said that a school careers counsellor or teacher (usually their mathematics or science teacher) had encouraged them to consider engineering, whilst two women said that their teachers had discouraged them from considering it. These studies all noted that most of the women in engineering had sought to find out about the profession before signing up and for them engineering was a conscious choice. In this they differed from the men who had some gender-related identification with the profession even if they were less than clear about what was actually involved.

McIlwee and Robinson (1992) found that another influence on women choosing to study engineering was that engineering was a good career choice. It was well paid and there were plenty of jobs. Although the same could be said of medicine (an alternative career for students who are good at mathematics and science), women who selected engineering did so because the training period was considerably shorter. These women perceived engineering as having an additional advantage over medicine as a career: it would be an easier career to combine with motherhood. However, the findings of McIlwee and Robinson in this regard have not been reflected as strongly in many other studies.

The main negative influence on women's orientation towards SET appears to be lack of interest in these areas of study and employment (Beder 1989; McIlwee and Robinson 1992; Anderson and Gilbride 2005; EWEC 2005). These studies found little evidence of interest in working in engineering among their female subjects at high school, and Beder (1989) explored how and why engineering was an unattractive study and career option to many women. Describing women as among ' . . . those who want to work with people rather than machines and numbers, those who care about social relations' (Beder 1989: 173), Beder found them to be discouraged by the heavy emphasis on mathematical and scientific aspects of engineering education and the neglect of the social and environmental issues. Another related finding in a recent US study (EWEC 2005) is that engineering is portrayed as being very difficult, with superior mathematics and science skills required. This was also endorsed by Metz's examination of professional society websites aimed at encouraging students to take up engineering (Metz 2007). The emphasis on mathematics and science and the difficulty of the degree are clearly not all there is to engineering study, but it can be enough to discourage women from considering engineering as a potential career, despite the advantages which are also described.

A further negative influence that relates to the concept of engagement is a troubling feature that has been frequently demonstrated: namely, that school students have little understanding of the work of engineers and are called upon to make choices based on inadequate information about which sort of engineering they could pursue. This situation immediately throws up a gender issue. Boys, who may not know much about what an engineer does, may well still recognize the profession as one appropriate for themselves. Many girls may be similarly unaware of what is involved, but they also know that it is not normally a line that girls follow. In other words a widespread ignorance about engineering, accompanied by the recognition of it being a non-traditional path for girls, may go some distance towards explaining their lack of engagement with it as a study choice (Gill, Mills et al. 2008).

South African studies by Le Grange et al. (1996) and Jawitz and Case (1998) provide yet another perspective on gender differences in interests in SET. Le Grange et al. found that male and female engineering students assigned different priorities to 15 global science-technology-society problems (though both groups agreed that the two problems with highest priority were population growth and fresh water supplies). Jawitz and Case's study ' . . . found that females, more often than males, named 'social identity' or making a contribution to the community or country, . . . as a primary draw to the field' (Jawitz and Case 1998: 190). The EWEC (2005) study found that it was important to stress the social issues aspect of engineering and the potential excitement of an engineering career in order to interest high school women.

Capacity: the Necessary Knowledge and Skills

McIlwee and Robinson noted, 'Excellence in math and science was a primary factor propelling our female respondents toward engineering. This distinguished them sharply from the men in our sample' (McIlwee and Robinson 1992: 24). In their study of South African engineering students, Jawitz and Case (1998) found that white females were mainly influenced to study engineering from their positive experience of mathematics and science at school. Beder (1989) reported from an Australian context the importance, for girls more than boys, of an interest in science and mathematics in choosing engineering as a career, and this has been confirmed by later Australian studies (Hobart et al. 2006; Mills et al. 2006). A clear majority of the women engineers interviewed in the study by Mills et al. (36 of 41) spontaneously identified their demonstrated ability in mathematics and science as part of their reasons for electing to study engineering. A feature of their school experience was that they took pride in their achievements in the areas of mathematics and science; they believed they were good at these subjects because they were naturally gifted. When asked about their decision to become an engineer, most spontaneously identified themselves as 'very good at', 'naturally interested in' and as 'having enjoyed' the mathematics and science subjects at school. This response held true across all types of engineering. It was evident that their aptitude for, and achievement in, mathematics and science formed a key factor in their identity formation as potential engineers. In contrast to the majority of the women, the men interviewed were much less likely to mention particular proficiency in these areas.

Metz (2007) reports that in the US there is virtually no difference between the performance in mathematics of 17-year-old males and females in the National Assessment of Educational Progress and that this has been the case for the past three decades. Even though their performance is similar, in many western countries there is a disparity in the number and percentage of females, compared with males, who choose to study the advanced mathematics and science subjects at upper school level that are required for entry to engineering study at university (Metz 2007).

Some previous studies have suggested that women in general have insufficient experience with scientific hobbies, or in playing with machines and gadgets in the early years, to encourage an inclination to pursue engineering (McIlwee and Robinson 1992; Evetts 1996; Frehill 1997). However, these and subsequent studies have also demonstrated that the women who actually do enter engineering often express a childhood fascination for, and an expert facility with, engineering technologies, rather than being deficient in this regard (Mills et al. 2006).

Continuity: Access to the Necessary Institutional Programs and Resources

Although many of the women who currently enter engineering do well in mathematics and science at school and express their affinity with it, these are not the only women that we should aim to attract to engineering degrees. From her industry experience, Beder (1989) challenges the necessity for the heavy emphasis on mathematics and science in engineering degree courses and the resulting barriers imposed on entry to these courses by requirements for good grades in these subjects. She points out that a woman's lack of the usual entry qualifications for these disciplines should not automatically be interpreted as indicating a lack of ability, but may have arisen from a lack of interest and motivation in studying these subjects as a result of the different socialization processes experienced by boys and girls. Bridging programs and alternative entry pathways that we discuss later are possible options to overcome these barriers to entry. It is also notable that in some countries (such as Malaysia for example; see Ahmad 1998) where students are given fewer subject choices at high school than in western countries, more women do study advanced level science and mathematics at school and this is reflected in higher percentages of women undertaking engineering at university.

Retention and Success of Women in Engineering Studies

Factors that influence women to stay and succeed in engineering study once they have chosen to enter it have also been the subject of many studies. One of the most extensive and often cited of these is that conducted by Seymour and Hewitt (1997) in the US. Their study examined the question of why some undergraduates chose to leave science and engineering study while others chose to stay. The study extended over three years and involved interviews with 460 students at seven institutions. The most notable finding of the study was that students who chose to leave science, mathematics or engineering degrees and switch to other majors were not different from those who persisted in terms of ability (grades) or in their discontent with their educational experience. A high proportion of all students found that teaching was poor and that it was difficult to get help with academic problems. However, those who left reached a decision over time that, combined with other factors such as a lack of identification with their potential future career in those fields, these problems were sufficient to cause them to leave (Seymour and Hewitt 1997). These findings were largely replicated in the later *Women's Experiences in College Engineering* project (WECE) (Goodman Research Group 2002) which used an online questionnaire to survey over 7,000 female engineering students at 53 institutions across the US each year from 1999 to 2001. They found that half of all students who left engineering 'cited dissatisfaction with their school's program (for example: grades, teaching, workload, pace) as a reason for leaving; another one-third mentioned their school climate: competition, lack of support, and discouraging faculty and peers' (Goodman Research Group 2002: v). Yet the majority of those who left had maintained good grades until at least the year before they left, but they were discouraged by their grades even if they were doing well. Such retention studies all stress the need to improve the teaching of engineering for all students, but also note that women in particular benefit from the presence of support networks and supportive environments, which may be provided by family members, male and female peers, male and female faculty members, targeted support activities or programs and external organizations.

Spotlighting

A vexed issue for many programs designed to attract and support women in engineering is that of enabling women to use such services without feeling as though they are being offered remedial assistance that may draw unwelcome attention to themselves. McLoughlin discussed these issues under the term 'spotlighting', which she defined as the 'singling out of women by gender in ways that make them uncomfortable' (McLoughlin 2005: 373). She noted three forms of this that ranged from intending to harm women (being overtly sexist) to the category relevant here: singling out women with the intention to help them. Her suggested solutions were the opening up of such programs to all students and targeting them more towards the 'pre-emption of non-professional behaviour' (McLoughlin 2005: 373). With the benefit of more than 30 years experience in these areas, we would certainly agree that the pre-emptive strategy should be a primary focus of universities, professional organizations and the corporate sector, as discussed further in Chapter 7. However, we would also strongly argue that opening the programs to all is not the solution. As mentioned earlier and expanded in Chapter 4, there are myriad factors at play which indicate that there is little further need to encourage or support males to enter and remain in engineering (beyond the academic support strategies that all universities should have in place for all students). However, as demonstrated previously, women cite support networks and encouragement as reasons why they remain in engineering (or the lack of them as reasons why they leave). Hence, even if some women choose not to participate in support activities (as is their right), others value them highly and so we suggest that these supports are vital components contributing to gender inclusive engineering education as a whole.

PROGRAMS AND INITIATIVES TO SUPPORT ACCESS OF WOMEN TO ENGINEERING EDUCATION

Universities, governments and professional organizations adopt a number of strategies designed to improve access to SET programs. Broadly, these strategies can be categorized as:

- Programs aimed at schools and teachers to improve awareness and increase recruitment, including:
  - activities in the schools
  - activities on university campuses
  - short residential courses
  - training and awareness programs for teachers

- Programs providing a link between school and university
- Programs within universities aimed at potential students, including:
  - o widening entry routes and procedures and special admissions schemes
  - o developing new programs designed to appeal to students from equity groups, such as double degrees
  - o scholarships for targeted groups
- Programs aimed at university faculty members.

There have been two recent publications, one in US and one in UK, which have summarized and evaluated a wide range of access and retention programs for women in engineering. In 2006 the National Academy of Sciences in the US published *To recruit and advance: Women students and faculty in US Science and Engineering*. This report provides a wide-ranging review of successful programs and actions being used at US colleges to recruit and retain women students and faculty members in science and engineering. Similarly in the UK, *Women in science, engineering and technology: three decades of UK initiatives* (Phipps 2008) examines over 150 initiatives to recruit, retain and advance women in science and engineering that have been run in the UK since the 1970s. In addition to the many examples described within these two publications, we provide further examples and discussion of access initiatives in each of the previously listed categories.

Recruitment and Awareness Programs: Targeting Schools and Teachers

There are two purposes in targeting schools and teachers. The first is to increase the pool of school students qualified to apply for university engineering programs, and the second is to influence the more senior school students who are already studying appropriate subjects, to actually apply for engineering programs at university. It is never too early to encourage female students to persist with science and mathematics throughout their secondary education. Consequently, some initiatives target primary and middle school students with the aim of exciting them about science and engineering. However, most concentrate on encouraging Year 9 and 10 students to persist with SET subjects until high school graduation, or target Year 11 and 12 students to encourage them to enrol in engineering degree programs.

Activities in Schools

Some argue that programs which mostly focus on middle or high school level students are too late to attract girls to engineering and are of interest only to girls who are already studying mathematics and science (Tomkinson et al. 2002). They therefore believe that it is important to focus on introducing the ideas of engineering to much younger students. One such program is the *Engineering Is Elementary* project which is part of the National Center for Technological Literacy at the Museum of Science, Boston (2004–2009). As of April 2009, this program is currently used in all 50 states of the US by over 15,000 teachers and over one million students. The program consists of a series of curriculum units developed for teachers to use at either Basic level (Years 1 and 2) or Advanced level (Years 3 to 5). Each unit focusses on a particular engineering field or science topic and has an accompanying storybook. One example is: *Catching the wind: designing windmills,* which considers the topics of wind and weather using a mechanical engineering theme and is accompanied by the storybook *Leif catches the wind (Denmark).* Another example, *An alarming idea: Designing alarm circuits,* focusses on the topic of electricity using an electrical engineering theme and is accompanied by the storybook *A reminder for Emily (Australia).* Each storybook has a child character who engineers a solution to a problem. Each character, either male or female, comes from a different country and the collection as a whole includes a range of racial backgrounds. All activities presented in the curriculum plan are gender neutral. The program also incorporates extensive research and evaluation on the effectiveness of the materials and related research questions (Museum of Science, Boston 2004–2009). In addition, the same organization has developed for grades 9 to 12 a year-long high school course called *Engineering the future: Science, technology and the design process*, which involves four projects with hands-on activities and related readings (Museum of Science n.d.). Both programs are supported by a large range of corporate and government sponsors.

At the University of South Australia an award-winning peer-mentoring program that links current engineering students with students in high school classrooms has been running since 2002. The program is managed by staff at the university but the primary interface is with electrical engineering undergraduate students who work for two hours per week on a volunteer basis with students in Years 10 to 12 on a range of projects related to robotics. The program has been supported throughout by a local electronics company but has also received significant funding support from state and local governments. It now involves over 25 schools and 1,500 students each year (UniSANews 2005).

Further strategies to encourage senior school students to apply for engineering programs include talks, booths at career fairs and investigative activities by engineers, using marketing devices which are designed to appeal to particular groups of students and collaborating with school staff to in order to encourage them to recommend engineering to students considering university studies. One aspect that must be considered in these activities is the importance of including women students as speakers. This provides visible concrete role models for female high school students.

Activities on university campuses

Many of the programs undertaken on university campuses have similar formats to the short visit activities undertaken at schools, but they have the additional advantage of exposing students to both the physical environment of engineering (such as visiting and using the laboratories) as well as the student environment (through meeting current students and faculty members). One issue that should be carefully considered is the selection mechanism used to determine which students, particularly females, are invited to attend such events. Hobart et al. (2006) found that some female students were not necessarily encouraged or offered the chance by their school to attend such programs. Students were frequently chosen by their schools for such opportunities based solely on their grades. While this may be the easiest way to choose students to be involved, the wide range of entrance score results for engineering students in that study showed that students who may be most interested in mathematics, science and engineering are not necessarily the very highest achievers. It is important that these programs are available to those with an expressed interest in pursuing engineering study, regardless of their scores, since there are alternative pathways open to them, as discussed in the next section.

Short Courses

Universities in several countries offer short courses for female students, frequently as summer residential or day camps, to introduce them to engineering and encourage them to consider careers in an engineering field. One example from Canada is the Ryerson University *Discover Engineering* summer camp (Anderson and Gilbride 2003), which has been running since 1991. The camp is targeted at female high school students who have completed Year 10, but students from other high school years are also accepted. It is a week-long day camp based on hands-on activities. The camp counsellors are female undergraduate engineering students, and the various sessions are taught by female professors and alumni. Exit surveys and evaluations have been conducted since the camp began and these have shown that on average, 80 per cent of camp participants went on to university study with over half of these enrolled in engineering. Of those enrolled in engineering, almost three-quarters said that 'the summer camp experience greatly or moderately influenced their decision' (Anderson and Gilbride 2003: 89). Ryerson University has since added a series of other activities to their outreach program, including a high school workshop program, a one-day career conference for high school girls and a one-day Girl Guide conference for girls aged 9 to 12 (Ryerson University 2009).

Another different but long-standing example of a short course type of intervention is the FEMME program developed by the New Jersey Institute of Technology (NJIT 2009), which has been running since 1981. The FEMME program is an intensive five-week summer course for girls who have completed Year 4 to Year 8 at school. The program for each year level is focussed on a different field of engineering, ranging through environmental, aeronautical, mechanical, chemical and biomedical engineering. Each program includes classroom discussions, laboratory experiments, projects and field trips. Detailed follow-up evaluations of the program were undertaken in 1987 and 1994 which showed that it was highly successful in encouraging the participants to continue with mathematics and science through high school as well as to enter science, mathematics or engineering majors at college (Cano et al. 2001).

Training and Awareness Programs for Teachers

Many of the school based activities incorporate a component of training and awareness-raising for teachers, but there are also some programs that focus very explicitly on teacher training. Ohio *Project Discovery* and the subsequent development of *Project Sustain* are one such example (Thom 2001; Ohio Resource Center 2009). The programs were funded between 1991 and the early 2000s and have been subsequently incorporated into state-funded centres for science and mathematics education. The original *Project Discovery* focussed on improving the knowledge and skills of middle school mathematics and science teachers through intensive summer courses, leadership programs and district workshops. Evaluations indicated that students of teachers who had undertaken the Discovery program performed consistently better in mathematics and science testing than students in non-Discovery teacher classrooms at the same schools and that this was particularly so for female students and African-American students (Thom 2001). In other words the program provided effective professional development for teachers in these areas. When National Science Federation funding for the project expired, the Ohio state government funded an extension of the program that shifted the focus to the undergraduate preparation of mathematics and science teachers through *Project Sustain*.

Other teacher training programs have focussed more specifically on gender equity training. One such program was the *Integrating Gender Equity and Reform* (InGEAR) project, which ran in a group of universities in Georgia from 1993 to 1998. It provided gender equity training for faculty involved in teacher education and focussed on encouraging gender equity among students, fair teaching practices and equal access to programs in science, engineering and mathematics (Thom 2001).

Programs Providing a Link Between School and University

Programs linking school and university fall into two main classifications. One is the type of program that takes students already qualified for entry to engineering study and offers an orientation opportunity for them to acclimatize to the university environment. Such a program focusses on exploring the campus, meeting future student colleagues, learning study skills and so on and may be called an orientation program. Most universities offer some kind of orientation activity for all students, but some will particularly target such activities towards groups such as women entering engineering. A good discussion of this type of program is provided by Watford (2007).

Many of the programs discussed earlier aim to keep women motivated to study the science and mathematics courses needed at high school to enable them to be eligible for entry to engineering programs at university level. However, there is also a need to provide access for other students (both male and female) who have not completed such studies but then wish to pursue engineering degrees. In many cases these students might not have graduated from high school but have pursued some kind of trade based career, or something entirely unrelated to engineering. Other students may have graduated from high school but not pursued science and mathematics courses to the level required for engineering degree entry. Programs developed to provide this opportunity form the second classification of those that link school to university and are often called bridging or foundation programs. One such program has been offered at the University of South Australia since the early 1990s. It is a one- or two-semester program that may be completed full-time or part-time and includes mathematics, physics and chemistry courses designed to bring students to the level equivalent to Year 12. Successful completion of the foundation program then allows admission to an engineering degree course (UniSA 2009c). Such programs, if successful, lead to a wider range of people in engineering programs and hence underscore the need for inclusive teaching.

Programs Within Universities

The previous discussion has focussed on programs that are directed towards school students and teachers, either within schools or at university campuses, as well as programs that are offered at university but do not provide degree qualifications. Another series of programs to provide access to engineering study that can specifically assist women students who might not otherwise enter engineering careers are provided within universities. These will be considered in the following three sections.

Widening Entry Routes

Widening entry routes is a strategy designed to remove some of the 'institutional' barriers to the entry of equity and diversity students to higher education, such as keeping strictly to standard selection procedures based on exam scores, even though some promising students may, through poor teaching, health problems, or economic necessity, have had to leave school early, or have otherwise had limited opportunities to develop their potential.

One such initiative offered at the University of South Australia and four other technology based universities in Australia is the Associate Degree in Engineering. This is a two-year program that includes the bridging mathematics and science courses combined with some of the first and second year engineering courses. Completion of the Associate degree gives one and a half years' credit into the bachelor of engineering program. What is quite unusual about the Associate Degree program is that one entry pathway is via an engineering aptitude test, a multiple choice test developed and conducted by the Australian Council for Education Research that 'assesses aptitude to think scientifically, solve quantitative problems, critically analyze information and display interpersonal understanding' (UniSA 2009a). The program is now in its second year and has been very successful in attracting a wide variety of students. Although it has the potential to be promoted more widely as an alternative pathway for women students, the current focus has been on students from the traditional equity groups which include rural based and low socio-economic groups.

New Award Programs

To address the issue of some women students finding engineering dull and lacking connection to society when it is presented in the traditional reductionist way, many universities offer double degree or double major programs which emphasize the relational and interdependent aspects of engineering with other disciplines, for example Engineering and Business, Engineering and International Studies or Engineering and Law. An extensive EU study called INDECS, conducted in 2001–2002, examined the attractiveness and acceptability of interdisciplinary degree programs to female students, faculty members and employers across Europe. It demonstrated that such programs appeared to be more attractive to women than the traditional single degree courses, especially if the interdisciplinary subjects start as early as possible in the curriculum (Womeng 2005). The American Society of Engineering Education (ASEE) also regularly reports the increasing adoption of liberal arts courses or subjects within or beside engineering degrees.

Other universities have found that even subtle changes such as the renaming and slight refocussing of traditional engineering degree programs (such as renaming 'chemical engineering' as 'pharmaceutical engineering') can attract significantly increased numbers of women.

Scholarships

Some universities offer scholarships specifically to women to attract them into engineering courses. This practice can give rise to challenges of discrimination from male students and faculty and must be carefully planned and developed in ways that preclude such negativity and ensure there is general approval for the project. At many places there are also scholarships available in engineering programs to all students.

PROGRAMS WITHIN UNIVERSITIES TO IMPROVE RETENTION AND SUCCESS OF WOMEN IN ENGINEERING EDUCATION

The previous section has focussed on a range of programs that seek to attract women to study engineering. But as we have seen from the earlier theoretical discussion, retaining women in engineering programs also requires significant attention, as well as ensuring their success. Such strategies designed to improve the confidence, satisfaction, retention and success of female students may be intended to support individual students or be aimed at changing the culture of the engineering departments (which will be discussed in Chapter 7).

A wealth of examples of initiatives aimed at individual students or groups of students are discussed in both NAS (2006) and Phipps (2008), as well as in the bibliographic references provided by Watford (2007). We will briefly examine a range of additional examples and issues relevant to such initiatives.

Programs Aimed at Students

Initiatives designed to support individual female students in succeeding and remaining in engineering programs can be further subdivided into the following groups:

- Support programs for individual women
- Specific courses and programs for women
- Support networks

Support Programs for Individual Women

Initiatives in this category may include mentoring activities, role models, career counselling, individual learning support and scholarships. Mentoring, role modelling and career counselling may be provided by either senior engineering students at the university or by faculty members. Mentoring programs may be formal or informal, face-to-face or electronic. There is an extensive literature on mentoring which is not discussed here but readers are referred to Watford (2007), Wasburn (2007) and Wasburn et al. (2008) for recent research on mentoring that is relevant to female engineering students. One issue that is often of concern with such programs (as with many women in engineering initiatives) is the unfair demands made on the time and effort of female students, and particularly female faculty members, when compared with their male colleagues, simply due to the small number of female mentors available. When this time commitment is not able to be sustained, the success of such programs can suffer. One such example was discussed by Murphy et al. (2007) in examining the reasons behind the success of the Industrial Engineering program at the University of Oklahoma in achieving parity of the sexes in its enrolment. Although in 2000 the student enrolment was 55 per cent female and the faculty composition was 40 per cent female (four of ten), by the end of the study in 2005 the female enrolment had dropped to 34 per cent, although this was still above national averages in that field. The two major reasons suggested for this change related to the loss of role models: senior, charismatic students who graduated and the highly visible female faculty members who had reduced their involvement and availability due to growing family commitments. As the authors noted, these changes might be small, but such 'small changes can have a large impact' (Murphy et al. 2007: 250).

Specific Courses and Programs for Women

Women-only classes or courses within universities have sometimes been provided as a means to allow female students to acquire practical skills, such as electrical circuit construction or computing skills, or to undertake difficult topics in smaller, more supportive environments. However, such practices are becoming less common, sometimes because women prefer not to be singled out from their male colleagues in such a way (see 'spotlighting', discussed previously) and sometimes due to the threat of litigation on the basis that the practice is discriminatory in countries such as the US (Wasburn 2007).

There are also a few women-only universities that offer engineering programs, such as Smith College in the US. An early European example of this was the École Polytechnique Féminine (now known as EPF) which was founded in 1925 by Marie-Louise Paris as a women-only engineering university. In 1994 it was required by the French government to become co-educational and the numbers of women graduating subsequently decreased. However, females still constitute 33 per cent of its student body, which is one of the highest percentages amongst French universities (EPF n.d.).

Support Networks

Although support networks for female engineering students may arise from some of the other initiatives already discussed, sometimes active interventions are used to develop these. One example is the residential community where female engineering students are assigned to the same residence hall or a single floor of a residence hall. Watford (2007) provides a comprehensive discussion of such programs. However, these are relevant only in countries where living on university campuses is common practice, as in the US, but would have no application to countries such as Australia where most universities do not have residential halls since students typically commute to campus.

A different type of support network has been developed at Tufts University in the US: the Nerd Girls (Loftus 2007). The group was founded in 1999 by Dr Karen Panetta and involves female students from the range of engineering disciplines at Tufts, who collaborate on renewable energy and community service projects and provide outreach programs to local schools. The Nerd Girls has now developed into a national and international group via their web presence (www.nerdgirls.com).

Programs Aimed at University Faculty Members

Clearly one strategy aimed at engineering faculty members is the introduction of gender inclusive curriculum into the courses and programs that they teach, which is the topic of the remaining chapters of this book. However, another strategy that should also form part of an institutional strategy to retain women engineering students is the provision of training in the equity and diversity issues of engineering education and the development of resources to support such training. This strategy has been adopted by a number of universities, such as Heriot-Watt University in Scotland; University of California, Davis; NECUSE (a consortium of New England universities) and Purdue University (Cronin et al. 1997). The resources developed are usually in the form of 'guidelines', videos and websites. Bianchini et al. (1999) identify professional development of teachers as the key to effecting change in the science curriculum in response to the issues raised by gender and ethnicity.

WOMEN IN ENGINEERING PROGRAMS

In many universities the role of coordinating all of these access and retention activities is the responsibility of a designated Women in Engineering Program, or Women in Engineering Officer. Between the late 1980s and early 1990s this strategy was adopted in many universities in Australia, UK and US for example, and it is no coincidence that this was the period which saw the greatest rate of growth of women engineering students (Holland 2007). Regrettably, as university funding has decreased, many of these WIE programs or officer positions have been discontinued, sometimes on the basis that numbers of women students were not increasing sufficiently to justify the expense, without apparently considering that the cultural change required is a slow process. Alternatively their functions have been 'mainstreamed' into Equity and Diversity units or engineering faculties or departments. In the vast majority of cases this means the loss of most, if not all, of the programs specifically targeted for women in engineering. In Australia, for example, where in the late 1980s to mid 1990s almost all universities had WIE programs, female enrolments in engineering more than doubled (Holland 2007). It seems unlikely to be a coincidence that in 2009 only three Australian universities (of 32 that offer engineering programs) still have dedicated WIE programs and female enrolments in engineering across Australia have been stagnant or declining for the past five years, as described in Chapter 2.

In the US the number of WIE programs grew from 10 in 1990 to 77 in 1998, but of those 77 only 45 had existed for three or more years and had paid directors with formalized activities (Knight and Cunningham 2004). The first and most long-standing WIE program in the US began at Purdue University in 1969, and for almost all of the past 30 years Purdue has graduated a higher percentage of female engineering students than the national average. The program currently offers the full range of initiatives that have been discussed in this chapter, including K–12 outreach activities on campus and in schools; a wide range of undergraduate support programs including mentoring, role modelling, learning support, female-only courses and residence halls; graduate student mentoring programs and diversity workshops with faculty members (Purdue University 2009c).

Although it is generally believed by women in engineering activists that WIE programs contribute to improved access and retention of female engineering students, the effects of such programs can be difficult to measure. In 2002 the *Women's Experiences in College Engineering* (WECE) project included an extensive evaluation of WIE programs in the US (Goodman Research Group 2002), examining the range of activities that they offered, surveying and interviewing directors of programs as well as students, faculty members and engineering deans about WIE programs. The WECE study found that women who stayed in engineering programs were more likely to participate in support activities, often provided by WIE programs, than those women who left. In addition, nearly one-third of students at engineering schools with WIE programs reported that this presence influenced their decision to attend that school. Hence the study concluded that WIE programs did meet many needs of female engineering students, but noted that the biggest challenge was the perception by some female students that they were providing remedial services and that participating in such programs would draw unwanted attention to themselves (the 'spotlighting' issue discussed earlier).

Both the WECE project (Goodman Research Group 2002) and a similar UK study (Cronin and Roger 1999) noted that one of the difficulties in determining whether a WIE program was successful or not was the different assumptions behind the programs and the consequently different strategies adopted by them at different institutions. A related study that focussed on in-depth interviews with 10 WIE program directors in the US looked at the characteristics of the most and least successful WIE programs (Fox et al. 2009). They noted that the least successful programs focussed more on helping individual women students cope, whereas the most successful programs provided that help but also focussed on institutional structures and trying to change them. The importance of institutional approaches is further discussed in Chapter 7.

GOVERNMENT, CORPORATE AND PROFESSIONAL SOCIETY INITIATIVES

The programs discussed so far have almost all involved a university as the service provider, although many programs rely on additional corporate or government funding support to operate. However, there is another range of both attraction and retention initiatives that are initiated and managed by government bodies, the corporate sector and engineering professional societies.

One notable example of an initiative that involved a significant partnership between government, the corporate sector and a group of universities was the establishment of five regional professorships for women in science and engineering across Canada in 1996. It came about as a recommendation from the Natural Sciences and Engineering Research Council of Canada (NSERC) Task Force, which required that the positions be open to women only and be spread geographically across the country. A condition of these professors' appointments was that each must take responsibility locally and (in collaboration with the others) nationally to move the women in SET agenda forward. Sadly this initiative arose out of the tragic Montreal Massacre in 1989, when a 25-year-old man who had unsuccessfully applied for admission to engineering at the University of Montreal, walked into the engineering school and fatally shot 14 female engineering students, deliberately targeting women and blaming them for his failure to gain admission (Zywno et al. 2000). The regional professorships are still in existence and still operate as a partnership between NSERC, corporate sponsors and universities (NSERC 2009).

Professional engineering societies in many countries organize programs for school students to interest them in studying engineering. These may be targeted at primary or secondary school students and are often conducted in conjunction with corporate sponsors. Examples include model bridge building competitions, science and engineering fairs (Engquest and the Siemen's Science Fair are Australian examples). Other programs targeted at school age women are sponsored by government funding such as the UK Women into Science, Engineering and Construction program (WISE).

In most countries professional women in engineering societies also exist in some form. Although these societies are primarily targeted at women already in the engineering profession, most also undertake outreach and support activities in schools and universities. A leading example is the Society of Women Engineers (SWE), based in the US, which has student sections on many campuses that often work closely with WIE programs where they are present. A similar group is the Women's Engineering Society in the UK, which was established in 1919. There is also an international organization that works in partnership with many of the national groups: the *International Network of Women Engineers and Scientists* (INWES). In other countries women's groups or committees are formed as subgroups of national engineering organizations such as the Women in Engineering committee of Engineers Australia. Whilst these groups all hold face-to-face meetings and conferences to allow networking and support of their members, the presence of resources on their websites as well as e-networks such as MentorNet also provide invaluable support to many women engineers and students. Such electronic resources also avoid the issue of 'spotlighting', which has been discussed earlier.

SUMMARY

This chapter has considered strategies designed to attract women to engineering study and then to help them succeed in their study and complete their engineering degrees. Many of the access strategies are focussed in school, some starting from as young as kindergarten age. Retention strategies in universities focus primarily on supporting the individual student, such as encouraging students to join, or form, networks; providing role models; and providing learning support and financial support in the form of targeted scholarships and fellowships. Other strategies aim to make the culture of engineering departments more inclusive of female students by widening access policies, providing dual degree programs and providing equity training for staff. Any or all of these strategies may be sponsored or managed by a range of groups, primarily universities who are trying to attract and retain women to their engineering programs, but also by governments, professional organizations and the corporate sector. Over the past 30 years such strategies have been implemented throughout the western world, and yet the gains in numbers of women in engineering have been relatively small and have now stalled in many countries. A large number of such strategies are tied to insecure funding or are heavily dependent on a small number of individuals and hence are vulnerable to discontinuation. For the same reasons very few of these programs have been rigorously evaluated with regard to the achievement of program goals (increasing numbers of women in engineering) as distinct from evaluation of participants' satisfaction with the program. This should not be seen as a criticism but rather as a consequence of the nature of the funding and support of the programs. Evaluations of the achievements of program goals require long-term data, potentially gathered over many years if the programs are targeted at early primary children, and usually neither the funding nor the staff continuity is available to enable this to be done.

All of the programs and initiatives discussed in this chapter are important and we heartily endorse their continuation, but we believe there are two significant elements that must also be addressed before progress can be made. One of these is the undergraduate curriculum. As mentioned in Chapter 1, only two of the 211 grants funded by the National Science Foundation in the US *Program for Women and Girls* between 1993 and 2003 directly addressed the engineering curriculum (NSF 2003). This is yet further evidence that for too long the engineering curriculum has been seen as inviolate and untouchable, rather than recognizing that it forms a very significant contribution to the problem that studying engineering remains unattractive to women and presents them with significant challenges in remaining engaged with their study if they do choose to attempt it (see quote at the beginning of Chapter 1). In Chapter 4 we examine the theoretical context of teaching and learning in an engineering environment as well as the theoretical background of gender inclusive curricula. We further develop this in Chapters 5 and 6 by providing practical advice to engineering faculty members and departments on designing and implementing such a curriculum in practice. The final crucial element in the picture, which we examine in Chapter 7, is the institutional approach which must be adopted to ensure that all of the strategies discussed in this chapter, along with the gender inclusive curriculum strategies in the following chapters, are embedded and supported within the culture of an institution to enable their success.

4 Learning, Curriculum and Gender
Theoretical Contexts

INTRODUCTION

In current educational theorizing, curriculum involves so much more than the setting out of topics to be taught and learned, the latter being what we used to know as 'syllabus', a term that gets little airplay nowadays. 'Curriculum' in current usage involves the content of teaching, the methodology of teaching and also the people involved: the what, how and who questions introduced in Chapter 1. In this book we argue that attention must be paid to the social composition of the class and the cultural background of the educator and students in the planning and practice of engineering education in order for it to qualify as inclusive. The understanding of learning outlined here explains the reason for this claim. At the same time, teaching, learning and assessment remain the core business of every formal educational endeavour, and so it is important to consider how these processes are understood if we are to make effective interventions. Any educational action is predicated on an understanding, sometimes implicit, about what it means to learn and, relatedly, what it means to teach. Consequently, we begin this chapter by looking at learning theory.

LEARNING THEORY

> Intelligence is what you use when you don't know what to do.
>
> *Jean Piaget*

Up until the twentieth century little attention was given to theories of learning. Learning was seen as operating by osmosis. Knowledge somehow seeped into the consciousness of the learners. People came to learning as passive empty heads waiting to be filled up. Thus teaching involved a good deal of exercises in repetition and memory, and those few people clever enough to make connections between processes and outcomes were seen as highly able. Those abstract thinkers who could see through to the principles of mathematical reasoning stood out from the majority of students, who simply relied on rote memory and guesswork. Those who were lucky enough to get right answers were seen as diligent but not necessarily intelligent. Schooling played a key filtering role in separating the clever ones from the others.

The early twentieth century saw a great deal of work on questions of intelligence, which was always ill defined but rested on a notion of innate ability and constituted a dimension around which the population could be measured and divided. Bright people, the highly intelligent, were seen as able to learn anything they turned to, whereas those at the other end of the spectrum were regarded as unintelligent and slow learners. Of course these ideas had little relevance for the practice of teaching, which went on in time-honoured didactic ways: the old chalk and talk. No matter what teachers did, the bright shone through and the others were left to languish.

The work of Jean Piaget challenged these simplistic notions with regard to intelligence and learning. Piaget's (1928, 1983) theory showed the learner (in his case the child, or *l'organisme,* as he originally wrote in French) as necessarily active in the learning process. This idea was extremely radical when it first appeared, but by the second half of the twentieth century it was broadly accepted by educators across the western world. Rather than having students sit in serried ranks waiting to be told, classrooms gradually became places in which students were expected to participate and to become engaged with the learning taking place. From this position teachers became much more than providers of information; now they were also responsible for creating an environment that facilitated student engagement. Learning came to be seen as no longer the preserve of the bright, highly able student but rather as an indicator of the quality of the teaching. Good teachers were recognized as promoting learning, and their classrooms were places in which everyone—not only the super bright—learned; thus these teachers were working inclusively.

While the active learner was the central tenet of Piaget's thesis, he had detailed qualitatively different developmental stages through which every learner must pass if she or he is to achieve the capacity to engage with abstract ideas and formal logic, that is, the sort of study that is expected of tertiary students. A good deal of educational research on elements of Piaget's propositions followed, and while there is still much debate about, for instance, the ages at which the different stages are typically achieved and whether the sorts of abstractions he talked about are more applicable to scientific knowledge than humanities based study, there remains a large degree of consensus about learning being an active process and that thinking becomes more sophisticated with increasing age.

Piaget's emphasis was on the internal working of the mind in terms of a structural progression. His theoretical framework formed the basis for the work of Russian psychologist Lev Vygotsky (1978), who added a crucial dimension to this body of work around learning. Vygotsky continued with the idea of the active learner but also insisted that the social context was critical to understanding the learning process. For Vygotsky, learning was necessarily a social process, not to be thought of as confined to the internal operations of one person's mind. In his view the active learner required the opportunity to work with the new knowledge, to think about it, discuss it and thus to engage with it in the company of other learners as well as teachers. From this perspective the social context of the classroom becomes an aid to effective learning and teaching. Learning is seen as a joint endeavour involving mutuality and support. Certainly most classrooms maintain the need for some direct instruction, but the teacher's role is to model the ways of thinking about the topic rather than to simply supply facts and methods to obtain the one right answer.

From the social context perspective, collaborative team work, a familiar feature of many engineering classrooms, is seen as one means of enhancing learning. Of course, this approach requires attention to group skills and processes and careful monitoring to ensure everyone has a chance to contribute and to preclude one or a few voices dominating. These conditions mirror almost exactly those written about by educational philosopher Jürgen Habermas (1987) in his formulation of the 'ideal speech situation' as a central feature of effective and inclusive learning and teaching. Vygotsky retained the central Piagetian idea that the learner actively builds her or his own learning, and this theory, or group of theories, led into the currently accepted learning theory: constructivism.

Constructivism

While there are many variants of constructivist approaches to teaching and learning, the central issues remain and can be identified as follows:

- Learning is an active process and requires the learner to engage with the learning, not just passively take it in or watch from the sidelines.
- All new learning constitutes a process of building on earlier knowledge; this is another derivation from Piaget's theory. So it becomes important for the teacher to establish what is already known and to think about what other experiences the student brings with him or her which might impact on learning. In this way the work for the teacher and the learner becomes that of making connections between the new knowledge and the material already known.
- There is no one way to learn. Some students may be more used to, or even more 'naturally inclined' to, a certain approach, but that does not mean they cannot cope with other ways. Learning is facilitated by a variety of approaches in terms of both teaching and learning. It follows that assessment should involve a range of different forms if it is to form an accurate indicator of the learning that has transpired. This point will be developed in a later chapter.
- Learners should be encouraged to become conscious of their learning, and teachers of their teaching.

In constructivist classrooms questioning is encouraged, along with space and time for discussion. Constructivist teaching does not mean doing away with lectures, but it does insist on the need for other learning spaces as well, in which students have the opportunity to work together on the material through discussion, joint projects, suggested problem setting and self evaluation of their learning. In addition, constructivism implies the necessity for learners to be able to reflect on their learning, to try out new ideas and to make mistakes as they engage in the intellectual work of assimilating new knowledge into their previous understandings. But this sort of education should not be seen as any minimalization of the teacher's role. If anything, the teacher is even more important in the constructivist classroom. Along with presenting material in a way that provokes discussion and allows for different approaches, the teacher must attend to the quality of feedback so that this aspect of teaching models the sort of engagement that promotes learning, even as it evaluates it. In a constructivist model teachers are expected to reflect on their teaching and to identify approaches that are seen to be productive of student engagement along with those that are not. Thus a constructivist approach implies a heightened awareness of teachers to the multiple factors that impact on student learning in any classroom. For present purposes the centrally important idea is that no student, and no groups of students, should be precluded from being actively engaged in the learning process—the hallmark of inclusive teaching and learning.

Constructivism has become the dominant learning paradigm in recent times and has many implications for inclusive curriculum work, which we elaborate in the next section.

CURRICULUM

As noted earlier, the term 'curriculum' is used here in its broadest sense, that is, covering all aspects of the teaching and learning experience to include assumptions about the prior experience and interests of the students; the syllabus (or content); the teaching and learning methods used; management of the classroom environment and the ways in which students are assessed. From a constructivist viewpoint the curriculum refers to:

- the knowledge seen as essential for the course, along with all prior learning which may impact on the learner's approach to the topic
- instruction and assessment strategies which allow for different learning approaches
- the socio-cultural environment in which students engage with that learning and where there are opportunities to discuss the topic, to ask questions and to volunteer opinions
- the teacher's modelling of the thought processes involved across a range of different approaches

- the energy and enthusiasm developed by the teacher and the group of learners
- individual levels of motivation and perspectives of relevance of learning.

Hence from a constructivist standpoint it is evident that the educator's task is much more complex than simply assembling key points in a particular knowledge area and presenting them to the learners to take in or not. From this perspective both educator and students are seen as mutually involved in a process through which learning is modelled, discussed, challenged, reiterated and assessed. This leads to an examination of the ways in which questions of gender might enter into this process of knowledge acquisition.

Gender and Curriculum

The particular focus of this book is on the ways in which gender operates as one dimension of the socio-cultural environment of teaching and learning in engineering. We do not imply that gender is the only variable that educators need to think about in their teaching, but we do suggest that it is a particularly important one. It is also true that until the past twenty years in most western countries gender was not seen to constitute an important consideration in engineering education, largely because of the homogenous nature of engineering student populations. As we have seen in Chapter 2 that situation has changed radically and looks to continue to do so with the advent of increasing numbers of engineering students from different cultures and the small but significant numbers of women. The argument about the need for more engineers has been developed in Chapter 1 and there is a widespread consensus throughout the developed and developing world that engineering skills are valued and important. So the quest for more engineers is strongly pursued. To the extent to which engineers of the future will work in social environments much more diverse than those of previous generations, it is imperative that engineering educators begin to prepare students for that increasing diversity. The first step is to examine the teaching and learning processes in engineering, that is, the engineering curriculum.

Few would deny that engineering is essentially an applied profession. Although engineering has been described as having difficulties with its public image, it is easy to point to the myriad ways in which engineers have contributed to the development of modern society. Many of the examples of engineering achievements are concerned with people's lives being made to run more smoothly, with more accessible communications and increasingly efficient systems to make daily life more comfortable. Since the effects of developments in engineering have ready application in our everyday lives, they become part of the normal, taken-for-granted living conditions and are not thought of as 'engineering'. The television works, the car runs along the road and across the bridge, the city buildings rise up to the sky and so on.

The reason for the claim that gender is a singularly important variable to take into account from a teaching and learning perspective arises precisely because gender forms another part of the taken-for-granted assumptions about the world in which we all live and share. This means that all of us carry unexamined gender assumptions which we unconsciously transmit in our every social interaction. However, when such assumptions become part of the teaching and learning process, they can negatively affect those who are made to feel like outsiders—that is, those who are not mainstream students, those less able to accommodate the material to be learned, those less able to speak out in class and those less likely to score high marks in the examinations. Because of the ordinariness of gendered thinking, even the students may be unaware of such discriminating processes beyond a general sense of feeling 'not right' in a particular class and not able to perform at their best level. We now turn to the evidence for this and related phenomena.

Research has demonstrated that the ways in which specific knowledge is taught tend to become tailored to the interests and perspectives of either the teacher or the dominant social or cultural group of students in the class, or both (Beder 1989; Lewis 1995; Goodman Research Group 2002). Since the majority of engineering faculty and students are male, it seems likely that male interests and styles of interaction will dominate, as indeed has been shown to happen. An Australian investigation of gender dynamics in an engineering classroom concluded:

> The most noticeable differences between female and male student experiences found in this study stemmed from four factors: their previous experiences; their learning approaches; the language used within the classroom and the lack of role models.
>
> (Burrowes 2001:1)

While recent research has challenged the once accepted notion of gender differences in approaches to learning (to be discussed later in this chapter), there have been few, if any, challenges to the claim of gender differences in prior learning experiences, classroom language and opportunity to interact with same-sex role models. From a constructivist standpoint, the atmosphere of the engineering classroom, the style of the lecturer, the language used and the images in the materials all add to the student 'building' a concept of who is rightfully an engineering student and who is not.

The language and materials used in engineering classrooms has been shown to teach implicitly that engineering is a masculine profession (Burrowes 2001). An early study of school science education focussed on the language used in mathematics and science texts and assessment exercises as evidence of the masculine bias in teaching and learning in those areas. Thus educators were alerted to the old textbook problems about men mowing meadows and tilers tiling bathrooms. Analysis of science texts also noted the predominance of men in white coats looking purposefully into test tubes in their pictures, all of which helped to convey the image of the scientist or mathematician as male (Lewis 1987). More recently Riley (2008) found that masculine pronouns and images still persist in engineering education literature despite the constant stream of advice to engineering faculty on this matter since the 1980s (for example Women in Engineering Committee 1999). Riley's response is to urge the importance of faculty members using inclusive language and avoiding stereotypes. She points out that consistently referring to engineers in examples and examination questions as 'he', and only mentioning or depicting women in stereotypical feminine roles or contexts, excludes women from a scientist or engineering identity.

Other contributors to what has been described as a 'chilly climate' (Hall and Sandler 1982) for female engineering students are erroneous assumptions by lecturers that all students have prior tinkering experience (practical familiarity with mechanical and electronic devices and appliances) (Lewis 1995); disruptive behaviour of majority groups (for example white male students throwing paper planes) (Lintern 1995; Jolly 1996); and an uncomfortable, sometimes intimidating classroom atmosphere because of racism, sexism, or similar attitudes (Mills and Ayre 2003).

A more recent study by Powell et al. (2007) found evidence that sexism persists in the following forms: sexist attitudes and humour, male lecturers struggling to relate to female students and higher visibility of female students. For example a female student felt singled out when her absence from a class was reported to her personal tutor, while male students' absences were not reported. Female students also commented that they received more help than the male students in the laboratories from lecturers and technicians, but there was no consensus between them as to whether this attention was welcome or not: some saw this as patronizing, while others welcomed it. Powell et al. comment that whether welcome or not, this is sexist because it implies an assumption that women are less capable than their male counterparts.

As female students are inevitably a minority in engineering classes, their marginalization in class processes may pass unremarked. After all, there are only a few of them. In these situations some female students may feel uncomfortable or excluded in class, with the result that they withdraw or are not as successful in their studies as they might otherwise be. This is an example of the concept of critical mass which has been associated with classrooms in which the minority group members gradually withdraw from the scene. While researchers debate the tipping point for critical mass, with most agreeing that anything less than 20 per cent can produce the effect, this is the constant reality for women engineering students. Their withdrawals and failures contribute to the ongoing popular impression that engineering is a difficult course for girls. The female student's self image is involved too. The struggle to claim a place on the non-traditional pathway, even for highly able and articulate young women, is not without costs. The discussion turns at this point to the question of identity.

IDENTITY

In considering the engineering curriculum from a constructivist standpoint we are alerted to the social realities of the typical engineering classroom in which there is a preponderance of male bodies. This gender imbalance immediately raises the question of the degree to which engineering as a field of study is masculinized, a claim frequently made in the literature. Hence we now consider how an understanding of the concept of identity can assist in achieving the aims of an inclusive curriculum.

Identity has multiple meanings. Once it was regarded, especially in the psychological literature, as a function of one's individual makeup or personality as though it was a product of innate proclivities in which gender was a dominant factor. More recently identity is conceived of as being produced by social experience and conditioning. In this view identity is conferred by social practice rather than being the product of some constellation of innate characteristics. This concept of identity is particularly useful to the concern with professional identity associated with engineering education. The popular impression of a particular educational or professional field can influence the ways in which it will be followed, identified with or be seen as more or less important by new entrants. Hence students' perceptions of their own identities can impact on the take-up of particular programs and the effectiveness of individual learning.

In current writing, identity is seen as comprising an individual's sense of self along with their capacity and willingness to embrace the set of expectations associated with a particular profession or position. This combination produces some of the controversial findings in investigations of identity among women engineering students in which some disavow any sense of discomfort in being a minority and reject any offers of special treatment, whereas others see their experience as comprising an ongoing struggle between their feminine selves and a masculine professional education (Mendick 2005; Osborne 2006). This bimodal effect has been found among investigations of women engineers (Stonyer 2002; Mills et al. 2006; Bastalich et al. 2007; Phipps 2008). Cross-cultural comparisons have shown there are cultural differences which interact with female engineering students seeing themselves as part of a minority (Godfroy-Genin and Pinault, 2006; Sagebiel and Dahmen, 2006). So just being a minority member does not in itself necessarily present as a disadvantage; what matters is the degree to which that membership is associated with less ability, lower self esteem, and different ways of engaging with the work.

For current purposes the point here is to suggest identity issues constitute another variant on the ways in which gender may become a feature of the learning environment in ways that position the women students apart from the men. This can occur in course content as well as in classroom treatments and processes and is another feature that can impact negatively on inclusive teaching and learning. Studies of the language of the engineering classroom, seen in the problems set, with women rarely pictured in the texts or mentioned in lists of high achievers in engineering, have concluded there is a bias in favour of males as the norm in engineering classrooms (Burrowes 2001). The ideal of the profession, a standard regularly referred to in all professional education, is similarly coded male. However our interest in this book is not so much to change the gendered association of the whole profession as to encourage engineering educators to be alert to the ways in which their teaching can work to confirm or to challenge the older stereotypes.

A key to success in science (and by implication in engineering), it is argued, is to assume or acquire the identity of a scientist (Bianchini et al. 2000; Hughes 2001). Researchers across many disciplines such as psychology, philosophy, sociology and women's studies, have explored how we define ourselves and who we are, as well as how we arrive at this understanding (Hughes 2001). While there is some debate between these different disciplines as to the main determinants of identity, we take the view that identity is a fluid concept and that we subjectively experience ourselves differently in different situations. For instance when considering one's own identity, individuals may adhere to a more or less fixed concept of 'who I am' in keeping with humanist notions of being true to oneself, holding to one's values and so on. At the same time the popular stereotypic assumptions about what people are like when they work in a particular field also affect those who aspire to enter that field or even think about entering that field. Professional identities are thus built up in terms of popular understandings along with the behaviour and attitudes of those already in the field. While identity is always inflected by gender, class and cultural background, it is the language used in the field through which the identity is continually modified that is the focus for the curriculum theorist. Hence the claim has arisen that we define ourselves through the way we speak and interact when occupying subject positions. From this standpoint the concept of identity is far from fixed but is rather a much more fluid idea that is modified and changed by ongoing experience (Hughes 2001; Allie et al. 2009).

Rather than taking on ready-made identities, some researchers suggest that we are the authors of ourselves. For example Bianchini et al. (2000) emphasize that individuals are active agents in the construction of their own identities and quote from Kondo (1990: 48) saying: '. . . crafting selves is an ongoing—indeed a lifelong—occupation . . . human beings create, construct, work on and enact their identities . . . '. And yet the pre-existing notions of, for example, engineer, scientist, dentist, do exist and do exert some influence over entrants to those professions. Our position on identity is to see it as a product of competing forces—those of the pre-given identity set connected with the profession and the individual's capacity, should they so choose, to do it differently. This will evidently be a site of struggle for those who through race, class, culture or gender experience themselves as different from the set of attributes commonly associated with the existing profession. An inclusive curriculum would have as one goal the idea of enlarging the ways in which the students understand engineering to include differences on all the dimensions listed. In this way the traditional masculine understanding of engineering identity is enlarged and enriched and its gender limited notion gradually overcome. The discussion now turns to the relationship between professional identity and academic success.

To some degree the association between different professions and identities is related to the idea of stereotype, namely the shorthand way of establishing what to expect of a person in a particular field. And so we speak of the absent-minded professor, the slick car salesman, the caring nurse and the go-getting executive. At the same time we know that these stereotypes do not hold true for every member of that particular field. Indeed the ones that do not conform stand out: the honest car salesperson, the singing detective, the unpleasant nurse. But we still use the shorthand, safe in the sharing of common knowledge even though we know it is not always true.

The professional identity of engineers has long suffered a similar narrowly defined description: the hard-hat and the Dilbert comic strip character for example. And of course it is coded male. 'Oh it's the cords!' cried one young woman recently when asked why it was that so few young women are drawn to engineering. By this she meant that the student engineers tend to be seen as unfashionable and poorly dressed (even among university students), as is easily observed in their choice of clothing such as shabby corduroy pants. Questions of image matter and are taken up differently by men and women in daily life. Not surprising then that they reappear in studies of women engineers (Waller 2004). Some women engineering students adopt a style that authorizes a feminine identity, whereas others choose to dress in a more gender neutral style.

Studies have repeatedly shown that the young women who go into engineering are aware that their choice is 'non-traditional', and so they tend to have carefully researched their choice and made it after much thought. By contrast the young men, often just as ignorant as the young women as to the actual content of the work, have streamed into engineering degrees knowing that they are on a regular and accepted pathway to an acceptable male career (McIlwee and Robinson 1992; Evetts 1996). Issues of professional identity have been shown to impact on the experiences of women engineers long after they have completed their studies and joined the workforce and at times these issues have been instrumental in their decision to quit (Bastalich et al. 2007).

In preparing for entry to a profession the idea of developing an appropriate identification is frequently stressed: just as the medical student is encouraged to don the white coat and have a stethoscope handy, so too does the aspiring engineer gather the symbols of professional belonging, both externally and internally. Being able to take on the role is seen as connected with academic success as well as success in later life. If acquiring the identity of an engineer is important to student success, the next question which arises is: what are the characteristics of an engineering identity? If we are continually re-constructing our own identities as a result of our life experiences, the extent to which our identities embrace or include those of a scientist or engineer will reflect our perception of the characteristics of these roles. We may also ask whether these characteristics fit more easily with some people than others.

There are some indications that engineers hold to particular ways of seeing the world and their professional place in it. For example, in a recent study Riley finds that: ' . . . engineers see themselves and the profession as apolitical, objective, and value-neutral' (Riley 2008: 109). In a rather similar vein, Hughes' research (2001) used critical discourse analysis to explore science students' perceptions of a scientist identity, and found that both males and females studying physical science regarded a scientist as abstract, rational and objective, valuing scientific knowledge principally for the status it brings. In this respect the scientist identity seemed very close to that of the engineers. However, some female subjects in the Hughes study, who were either about to drop out of science or studying biological sciences, took a different view. They valued relevance and human interest more highly than the abstract rational objective scientist attributes previously mentioned. Hughes does not suggest that all males and females divide along these lines, only that while most males held to one view of science and scientists a significant group of females did not. It is important to note that some of the women did agree with the identity expressed by most of the males. For example a female physics student from an immigrant Vietnamese family now living and studying in the UK acknowledged that as a result of parental pressure coupled with her own ambition: ' . . . you wanna do better than the people who live there so that you can show that you [are] like worthy to be there' (Hughes 2001: 282). In this she differed from the group of female students who rejected or queried a scientist identity. For this student, ethnic identity overruled the (non-)scientist subjectivities displayed by the other female students.

Hughes' point is that widening perceptions of the identity of a scientist can result in a broader constituency of potential scientists, to include more women and other underrepresented groups. As she says:

> While rigid and objectivist science is only compatible with a narrow range of student gender and ethnic identities, socially relevant and more constructivist science can generate a wide range of scientist

> subjectivities, increase the possibilities for scientist identities and thus open the way to a more inclusive science curriculum.
>
> (Hughes 2001: 288)

Another dimension of the role of identity in science and engineering education is the impact of a teacher's identity on the learning experience. Bianchini et al. report a need for science teachers ' . . . to carefully and critically examine their personal identities in interaction with their professional experiences' (Bianchini et al. 2000: 538). In their view insight into the perspectives of their students is important in promoting inclusion. Science teachers need to be aware of the ways in which gender and ethnicity shape scientific norms and practices, a view echoed by Riley (2008) in her call for engineering faculty to be more critical of curriculum content which may not interest minority groups or which privileges dominant groups. As an example of the latter Riley cites the US space program as being underpinned by the legitimacy of global dominance of western societies.

In summary, researchers conclude that the question of identity poses more challenge for women engineering students than for the men. In one researcher's terms, 'The masculine culture in engineering communities of practice involves more effort in identity management for women students than for their male peers' (Du 2006: 35). In this environment it is surely important for engineering educators to ensure they do not contribute to the identity management issues for women students by endorsing the view of engineering as a masculine domain.

Self Concept

Within the multiple ways of thinking about identity, the question of self concept is singularly important. How a person sees themselves as autonomous, worthy, capable or not is obviously going to impact on their behaviour, their life choices and the person they become. Studies of women engineers have consistently shown them to have a lower self concept than their male peers, indicating they feel less sure of their abilities, more nervous about their success in examinations and more daunted by the prospect of a new challenge. Studies in the 1990s found that women engineering students typically scored lower than their male peers on measures of self efficacy and self confidence (Felder et al. 1995; Brainard and Carlin 1998). Even more disturbing was the finding of a recent study in which women students' lower self efficacy was seen as relating to the finding of 'a significant decrease on feelings of inclusion' (Marra et al. 2009: 27) experienced by women engineering students at five different US institutions during their course of study. These findings were linked to previous studies showing that women students report receiving negative messages about their presence in engineering (Seymour and Hewitt 1997) through the language of the classroom and the images and people appearing in the texts. The significance of this work for the present project on gender inclusive curriculum is that it forges a definite link between women's learning outcomes and classroom processes.

Another American study in the same year found no statistically significant gender difference in most of the self efficacy measures (Concannon and Barrow 2009). However, there was one measure showing an interactive effect between coping and self efficacy, with the females reporting lower on the coping dimension. Interestingly the gender difference that emerged in this study related to the women being much more likely to have somebody to turn to for encouragement and support (63.8 per cent of females compared with 38.2 per cent of males), a finding that supports the idea that support structures and mentoring may be more salient for minority students and women. A related study found that women students were more aware of the context of the classroom and the people in it than were men and hence were more likely to be influenced by it (Kilgore et al. 2007). Clearly the designers of curriculum in engineering should take notice of the social settings if their teaching is to work inclusively. Noting that self efficacy was shown to be the best predictor of overall success in engineering, Vogt et al. concluded ' . . . classroom interventions need to be targeted at enhancing student self efficacy for success in engineering programs . . . ' (Vogt et al. 2007: 361), making the issue of direct concern for curriculum and practitioners.

In the interests of improving women's self concept and related self efficacy, engineering education should look to the content of curriculum in terms of rendering it more inclusive, more likely to picture women and men equally as confronting engineering tasks and more conscious of the social changes that have brought more women and minorities in the classroom. It seems likely that the provision of mentors and the structuring of course experience, whereby such mentoring was readily available and not limited to those who are struggling with their studies, would also help women students gain a stronger sense of entitlement within their chosen area of study.

LEARNING STYLES—ARE THERE GENDER DIFFERENCES?

Early research sought to relate the under-representation of women in science and engineering education and employment to gender differences in the skills and attitudes needed to learn effectively. Coupled with the traditional lack of mathematical knowledge in the senior school years, it appeared that girls and women lacked the capacity to study in those areas at tertiary level. In this section we look at the question of sex differences in cognition and in approaches to learning, or what is often termed 'learning styles'.

In their comprehensive analysis of sex differences in cognition, American researchers in the 1970s concluded that the differences between boys and girls were so slight and unstable as to warrant extreme caution in making the claim that they exist (Maccoby and Jacklin 1974). At the same time a British researcher conducted a similar overview and agreed with the American findings (Fairweather 1976). Fairweather added that differences in cognition were least likely to appear in the very young, giving a strong indication that differences that do occur arise from socialization rather than from innate physiological difference. However, in view of what were at the time persistent gender differences in attitude to, and achievement in, mathematics, these same American researchers attempted to identify the causes by looking for 'affective' (or 'attitudinal') variables, where 'affect' and 'attitudes' can be defined as 'beliefs about oneself and mathematics' (Hyde et al. 1990: 301). They found gender differences in two variables in particular: firstly, perceptions of the usefulness of mathematics, and secondly, students' levels of confidence in learning mathematics. In both cases males were found to be more positive than females. It is probably inevitable that findings of this nature eventually became grouped together as 'deficit' models of women's learning capacity. More recently, work by Rogers, an Australian neuroscientist, demonstrated that the differences in terms of achievement in mathematics had markedly diminished over the last two decades of the twentieth century (Rogers 2002). Her suggestion was that as the culture had become more gender inclusive and more women appeared in public life and paid work, it became increasingly acceptable for girls to be recognized as intelligent, and so the gender differences in achievement disappeared. By 2008 Hyde's team had conducted meta-analyzes of achievement in mathematics and was able to show that Rogers' prediction had been realized. The earlier regularly appearing gender differences in mathematics achievement had disappeared (Hyde et al. 2008). Although Hyde's findings were widely reported, changes in widespread attitudes, especially those about sex differences in style and capacity, are not easy to shift. As seen in the previous section, studies continue to find lower female self confidence in academic achievement, especially in the areas of mathematics and science.

As noted earlier, attitudes comprise an important factor impacting on learning. Male students continue to see mathematics as a male domain, even when women students do not (Hyde et al. 1990). Women studying in science and engineering identified a number of barriers to their success including ' . . . losing interest, intimidation, lack of self-confidence, poor advising, and not being accepted in their department' (Brainard and Carlin 1998: 9).

For some time educational research into cognitive factors had tended to disregard questions of the ways in which student progress is affected by their attitudes, previous training and the ways in which they engage with new learning. However, with the advent of constructivist approaches to learning, this separation between affect and cognition has increasingly been seen as unacceptable. There is now general agreement among learning theorists that social factors play a significant role in cognition.

The debate as to whether there are male/female differences in ways of knowing or learning has been subsumed in the learning styles debate, where the term 'learning styles' is used to describe differences in the ways individuals or members of cultural groups prefer to receive, process and present information and ideas. Some people, for example, find it easier to understand a new concept by reading a textbook, whilst others prefer a verbal explanation. Likewise, people may vary in how they most effectively demonstrate their understanding: graphically, verbally or in writing. A number of studies in the 1980s and 1990s claimed to have found gender and cultural differences in learning styles (Knowles 1990; Moxham and Roberts 1995; University of Western Australia 1996) and consequently drew significant implications for effective teaching and learning. Subsequent research has challenged these findings (Marra et al. 2009) and suggests that it is dangerous to assume such gender differences in working with students of either sex. Ingram and Parker had earlier warned of the danger of limiting 'the range of interactional strategies open to individuals' (Ingram and Parker 2002:33), a position we endorse.

Some of the most common claims about preferred learning styles in the literature on women in engineering education are that women prefer to work collaboratively, do best in group work and need a warm and supportive social context in which to shine. Several recent studies have challenged these general claims. In their study of modes of collaboration in an engineering classroom, Ingram and Parker (2002) showed that men and women alike displayed both gender-linked and non-gender-linked behaviour and perhaps more importantly, collaboration was less influenced by gender and more by such factors as a strong work ethic, team commitment and effective leadership (Ingram and Parker 2002: 33). Similarly Marra and Bogue (2008) studied classroom learning environments in engineering education and found no support for the commonly held view that there are gender differences in the ways in which teaching and learning are experienced. Moreover, other researchers have noted that there is no evidence to support the claim that matching teaching approaches to the learning style of the learner will have any effect on their learning (Litzinger et al. 2007). At this stage it seems likely that debates around learning styles will continue but in a modified way from the earlier theorizations. Where once gender differences in learning styles were hailed as providing a plausible explanation for the ongoing gender differences in educational outcomes (such as the small numbers of women qualifying as engineers), it seems now there is more caution in embracing this explanation. Learning styles, now described as 'fluid traits' even among the keenest proponents, are consistent neither across subject area nor across individuals and so hardly able to sustain the weight of gender difference explanation they once carried.

Thus it would appear that while most groups will exhibit a range of learning styles and some of these may be connected to gender, the relationship is not determinate. That is, it is not true to think that all men think this way and all women another, even though there may be some justification for saying that most males will adopt one practice and most females a different one. More likely, when the learners share the same discipline base, there is a high degree of commonality about their approach to learning as consistent with their study area (Marra and Bogue 2008). It seems there is more overlap between male and female approaches than previously thought. Also, and importantly, most people are quite capable of adopting non-habitual learning styles as they continue to learn. As we have argued earlier, a more flexible approach to teaching and learning, typical of a constructivist education, will benefit all learners—and teachers too.

However, the topic of learning styles is valuable in a consideration of educational reform in engineering. Firstly, it is useful for students, in the process of developing consciousness of their own learning as recommended by constructivist theory, to be introduced to the evidence that individuals have different learning preferences and to be helped to identify their own. As Felder emphasizes in the advice he gives students who have used his learning inventory to identify their own learning style profile, in their studies they will encounter some learning experiences which noticeably facilitate their learning, but they will find others more demanding. Learners need to work at the learning tasks they find more challenging, with the ultimate aim of being able to learn efficiently in any learning situation (Felder and Soloman n.d.). Secondly, there is often a mismatch between the wide range of engineering students' learning styles and the usually narrower range of faculty teaching styles, with many engineering faculty assuming that all students adopt (or should adopt) a uniform learning style (Felder 1993, 1996; Holt and Solomon 1996). The disadvantages for those students whose learning styles do not match the prevailing teaching style extend to their being less successful in the assessment process (Heywood 2005). Thirdly, there is evidence that learning improves when attempts have been made to expand the opportunities for students to employ diverse learning styles (Felder and Brent 2005).

Thus, one of the advantages of engaging with the debate about learning styles relates to its capacity to alert students to the possibility that they may be able to overcome a problem they encounter in their studies by attempting a different learning approach. Another advantage is alerting engineering educators to the existence of a range of learning styles and to the benefits of incorporating a variety of approaches into their teaching and assessment processes (Mills and Ayre 2003; Heywood 2005).

LINKING LEARNING THEORY TO THE GENDER INCLUSIVE CURRICULUM

We have linked learning theories and associated concepts to some issues of gender in engineering education. The principles of constructivism have been outlined, and the topics of identity, professional identity, self concept and learning styles have also been canvassed to allow further identification with some of the issues that may serve to exclude women and minorities in engineering education. We now explain and define the current concept of a gender inclusive curriculum in science and engineering and then examine the international evidence for its effectiveness.

The term 'gender inclusive curriculum' is used for a curriculum which has been consciously designed to recognize and acknowledge the evidence that, as well as all individuals bringing different cultural 'baggage' to their learning experience, there may be some distinctions between the baggage brought by males and females due to gender socialization. Such baggage will include prior knowledge, interests, approaches to learning and strengths in types of assessment tasks. In an applied area such as engineering, an inclusive curriculum recognizes that student experience of the applications of engineering in daily lives will vary according to gender, race, culture and class. If, as constructivist learning theory suggests, all new knowledge must be linked to the old, it is very important for effective learning that a student's teaching and learning experience makes reference to the different cultural elements that students bring. Importantly, it has been shown that all students, male as well as female, black as well as white, young as well as mature-age, benefit in terms of improved motivation and success from a flexible approach to teaching and learning (for example Kramer-Koehler et al. 1995; Tucker 2000; Fromm 2003).

Before moving on to the consideration of gender inclusive curriculum in today's engineering education, we start with a brief outline of the historical development of the concept. We believe this approach is useful to the educator who is interested in adopting the gender inclusive approaches and practices we recommend, since current research shows that the rationales for the earlier versions of the concept are still valid, although regarded today as incomplete. The issues for women in engineering education persist, and hence the strategies recommended remain necessary. In practice therefore, the gender inclusive educator needs to constantly monitor and re-examine his or her practice, in a process described by Rosser (1995) as a spiral.

Historical Context

Historically, the first approach to teaching students who differed from the dominant social or cultural group in a class was to ensure that these 'different' students were included, as opposed to excluded, in lectures and laboratories, giving rise to the term 'inclusive' teaching. Faculty members were urged to examine their teaching practices and ensure that in the assumptions they made about students' backgrounds, the examples and metaphors they used, and in the attention they gave to individual students, they were not excluding those students who came from cultures which differed from the dominant culture represented in the class (usually young, white and male) (Fennema and Leder 1990). Measures relating to the physical environment were also often suggested, for example lowering the level of laboratory equipment so that it could be reached by students other than six-foot-tall males.

In the 1990s several researchers and champions of gender inclusive curriculum in science, engineering and technology (SET) classified this evolution of attitudes towards minority groups of students along a spectrum or in stages: from 'not being noted', to 'acknowledged', 'accommodated', 'assimilated' and 'celebrated' (Warren 1989; Rosser 1995; Moxham and Roberts 1995; Willis 1996). Although these 'stage models' differed in detail, they all began with recognition of the heavily masculine orientation of traditional SET curriculum and moved towards enhancing the student learning experience in terms of process and content to become as fully gender inclusive as possible. Practical tips on how to be inclusive in all the stages are built into these models. In Table 4.1 we provide an earlier adaptation of Rosser's stage models that we developed to assist engineering faculty in our own institution to implement inclusive curricula. The value of these models lay in their usefulness as an introduction to the components of the gender inclusive curriculum for those to whom the concept was relatively new. By helping to locate one's own understanding of the gender inclusive curriculum along such a spectrum, the next stage of implementation was suggested, together with advice on how to do so.

In parallel with the stage models of the gender inclusive curriculum, feminist philosophers of science such as Bleier (1984), Harding (1991), Wertheim (1995) and Fox Keller (1996) questioned the validity of sciences developed and interpreted almost exclusively by white men. They argued that science which develops from these origins is necessarily incomplete and suggested, for example, that some critical environmental issues have been neglected because white male scientists with a predominantly industrial perspective have until now determined the priorities for scientific enquiry and allocated the requisite funding. Scientific research would have taken different directions, it was argued, if women (or indigenous people, or non-westerners) had been influential in these decisions (Webb 1995).

In the 2000s, several additional dimensions have been added to the gender inclusive curriculum spectrum that acknowledges, accommodates, assimilates and celebrates women in the engineering classroom. Science and engineering faculty are now also urged to be aware of their own and their students' identities when designing learning experiences and to raise awareness of gender discrimination in the workplace together with the skills to combat it (Hughes 2001; Riley 2008; Gill, Sharp et al. 2008). Somewhat more controversially, some educators argue that a gender inclusive curriculum also involves encouraging students to challenge the power of the curriculum to support social injustice; otherwise they may be permanently excluded from the benefits of society (Willis 1996; Bianchini et al. 1999; Phipps 2008; Riley 2008). Thus the curriculum itself becomes a focus for discussion, reflection and challenge.

Table 4.1 Stages of Curriculum Transformation

Stage	Title	Characteristics	Strategies to Achieve Change
1	Absence of women is not noted	Assuming that the objectivity of engineering renders it immune to gender considerations	Acknowledge gender influence and seek ways to incorporate them appropriately in the curriculum
2	Recognition that most engineers are male and that engineering may reflect a masculine perspective	Viewing the world from a male perspective	Explore issues of social concern. Set open-ended investigative-type problems
3	Identification of barriers that prevent women from entering engineering	Exploring why women are not attracted to studying engineering; how to attract them and how to reduce barriers at entry	Consider the learning environment as well as entry issues; remove the 'chilly climate' experienced by many women students
4	Search for women engineers and their unique contributions	Including the contributions of women engineers and discussing why they have often been 'lost'	Teach in cooperative and interdisciplinary ways. Discuss the social benefits of technological progress
5	Engineering done by feminists and women	Accepting and incorporating women's different perspectives and 'ways of knowing' in the study of engineering	Encourage development of theories and hypotheses that are relational, interdependent and multicasual. Use qualitative and quantitative methods in data gathering
6	Engineering redefined and reconstructed to include all people	Incorporating Stages 1 through 5 into a transformed inclusive mainstream curriculum	Base the philosophy, aims, objectives, content and way in which the curriculum is delivered on the principles of inclusivity

Adapted from Rosser 1995: 4–17.

Gender Inclusive Curriculum in the Context of Today's Engineering Education

The concept of a gender inclusive curriculum has evolved to include curricular innovations as well as pedagogical ones. That is, the curriculum content (or syllabus), as well as the teaching methods employed, should acknowledge the interests, attitudes and perspectives of the non-dominant groups in the classroom, as well as the dominant ones. The modern concept of a gender inclusive curriculum is therefore complex. It encompasses a spectrum of strategies and philosophies which, put very simply, will start with the relatively straightforward task of ensuring that there is no sexist behaviour in the class. A variety of teaching and learning methods will be used, and the course content will include social and emancipatory elements as well as technical matter.

Incorporating any of these approaches will improve the gender inclusivity of a course or program and the ideal aim is to incorporate all of them. To be fully inclusive, a curriculum will be inclusive in all its components. However, the extent to which individual faculty members can change or influence these curriculum components varies according to their level of responsibility. All teaching faculty have the responsibility for control of the classroom environment, subject perhaps to general guidelines. Teaching and assessment methods are specified in outline in the course approval documents, but again each faculty member has considerable discretion about how these are interpreted and applied in the classroom or laboratory. However, the course aims and objectives and content are determined at a more senior level. The classroom teacher must adhere strictly to these as set out in the course documents, although some flexibility is usually allowed in the examples and applications used. Significant changes to these components can only be made through formal processes, needing approval from course and department managers and, in engineering, usually from the profession too, through accreditation processes.

To elaborate on the concept of a gender inclusive curriculum in a way that is useful for both course coordinators and classroom teaching faculty, we present research evidence here as it addresses the three principal curriculum components: classroom management; teaching, learning and assessment; and course content.

Gender Inclusive Classroom Management

An early approach to gender inclusive teaching was to ensure that female students were not excluded from the learning activities in the classroom or laboratory. A study by Cooney (1991) of gender dynamics in her electrical engineering technology laboratory confirmed earlier findings in the UK, US and Australian high schools (Whyte 1986; Butler Kahle et al. 1993) that males tended to be the ones who handled the equipment, while the women students observed and recorded the results. Similarly, Rosser and Kelly (1994) in a study of nine US campuses found that males spoke up in science classrooms far more than females and received the most attention from instructors, despite the instructors believing that they gave equal attention to both males and females. This issue is still important, and the question of teacher time spent with some students rather than others continues to feature in the research on equitable teaching practices. Chetcuti, for example, urges school science teachers to be aware of ' . . . differences that can lead [to students] being marginalized and left out of the science learning community' (Chetcuti 2009: 95). At this point we remind readers about the importance of ensuring that the language and materials used in engineering education do not assume or imply that engineering is a masculine activity or profession.

Gender Inclusive Teaching and Learning

Earlier in this chapter we have demonstrated that the constructivist theory of learning provides a framework for effective teaching and learning. Students must be actively engaged in the process of learning; they build on what they already know, and so it is important to be aware that diverse backgrounds will result in diverse prior experience and knowledge. We point out that when classrooms become more diverse with the addition of women and students of other ethnic origins, the teaching and learning methods used must become more varied; otherwise the original group will remain dominant and privileged. We have looked at the debates about identity, self efficacy and learning styles through a gender lens and have concluded that the value of these studies in the engineering classroom is to raise the awareness of instructors to the need to build in a variety of learning experiences in order to encourage the participation and motivation of students with varied backgrounds, attitudes and perspectives. Furthermore, introducing students to the different attitudes, points of view and priorities of different social and cultural groups towards science and engineering issues improves the quality of education for all groups by encouraging debate and analysis.

As previously described, this emphasis on the value of a range of teaching and learning methods is not only an inclusive issue. While active and cooperative learning experiences are inclusive, they are also recommended as strategies to improve the learning of all students by, for example, Heywood (2005) in his comprehensive volume of research and development in engineering education across the world.

Gender Inclusive Curriculum Content

There are at least three approaches to making the curriculum content more inclusive. One is to use applications and illustrations which span a wide range of experiences and interests, for example considering the aerodynamics of dolphins as they swim rather than always using examples of formula one cars or aircraft wings in fluid mechanics courses. The second approach is to provide students with opportunities to investigate the social and humanitarian aspects of engineering as well the technical contexts. And thirdly is the opportunity to explore ' . . . how issues of gender and ethnicity shape scientific norms, ideas, and practices; and to what depth such issues permeate the scientific enterprise' (Bianchini et al. 1999: 73). Within this third approach we would also include the need to incorporate the discussion of issues faced by women engineers in the professional workforce, to encourage all graduates to consider how gender issues also permeate engineering and may affect their future careers (Gill, Mills et al. 2008). We explore all of these approaches in much greater practical detail in Chapters 5 and 6, but provide some additional theoretical context here.

In relation to the second approach, engineering faculty can no longer claim, as was commonly the case 25 years ago, that it is not possible to include non-technical content in the curriculum as this will never be accepted by the accrediting bodies. On the contrary, largely in response to industry pressure for engineering curricula which prepare students appropriately for the changing employment market, the accrediting bodies are increasingly requiring curriculum content beyond the technical, scientific and mathematical (Heywood 2005). For example, six of the 11 program outcome requirements for accreditation by the US Accreditation Board for Engineering and Technology in 2009–2010 involve non-technical skills such as teamwork, communication and ethics (ABET 2008; also refer to Appendix A). The program outcomes specified by the corresponding accreditation bodies in other countries, such as the UK (Engineering Council, UK 2007) and Australia (Engineers Australia 2008a) contain similar requirements. All of these topics fit well with the inclusive curriculum features suggested by theories of effective learning and the concepts of engineering identity and students' self efficacy.

We now consider the more controversial view encompassed within the third approach: that students should be encouraged to question the nature of scientific discovery in the context of who is privileged by it and who is excluded, a philosophy sometimes described as 'emancipatory' or 'liberative' (Willis 1996; Hughes 2001; Pawley 2004; Riley 2008). A higher education which encourages students to challenge, question, analyze and interrogate provides them with the skills to engage in the knowledge-building enterprise. Only by using these techniques, it is argued, can knowledge advance and applications which benefit humankind develop. Riley (2008), for example, argues persuasively that the positivist mindset which underpins most engineering education leads to uncritical acceptance of authority and deprives the engineer of responsibility for the consequences of technology. An example of her second year *Engineering Thermodynamics* course taught using 'liberative pedagogies' is given in Chapter 6.

HOW EFFECTIVE IS AN INCLUSIVE CURRICULUM?

We have presented the advantage of a gender inclusive curriculum in engineering as being the provision of a learning experience in which the minority of women students in a predominantly male environment is recognized by the teaching faculty. The curriculum is consequently reviewed to ensure that women are not excluded from any of the learning experiences, their full participation is encouraged and facilitated and they are prepared for the professional workforce where the masculine ethos is often stronger than in the educational environment. Possible consequences of a curriculum that is not gender inclusive are that female students are either disadvantaged, or discouraged and demotivated, and either leave engineering programs or fail to achieve the level of success to which they may once have aspired. The question arises as to how we can know whether these aims are being achieved, that is, whether female students have been aware of any discomfort, disadvantages or discrimination in the male-dominated environment of an engineering department and whether there is any evidence that a gender inclusive curriculum has improved retention, progression and success rates of female students.

Evidence of students' course experiences from a gender perspective has been sought by many of the faculty who have revised their courses to make them more inclusive, with positive results. Most of the inclusive curriculum exemplars in Chapter 6 report improvements in student satisfaction, and/or retention and success. Some other studies are cited in the following paragraph.

Kramer-Koehler et al. (1995) reported improved performance by women students at the Polytechnic University, New York, following the introduction of a new engineering freshman core course based on inclusive principles in 1993. The class was very diverse in terms of ethnic origins and was 17 per cent female. There was a strong emphasis in the curriculum on cooperative learning, 'hands-on' laboratory based activities, computer skills and oral and written communication skills, while mathematics was developed on a 'need-to-know' basis. The success of this course was evaluated against students' age, sex, ethnicity and school background, finding that '... women and/or Black and Hispanic students are significantly more enthusiastic about cooperative learning, hands-on learning, and peer-tutoring; and less enthusiastic about 'traditional' lecture-based teaching and learning' (Kramer-Koehler et al. 1995: 4a2.5). Importantly, it was also found that *most* students preferred the new approach and that overall retention rates improved by 50 per cent. Similar results are reported by Froyd and Ohland (2005) from Texas A&M University, after the introduction in 1994 of a new integrated first year program for some engineering students. The retention rate for women on this program was 72 per cent compared with 66 per cent for first year women students enrolled in other engineering programs. The same authors report from a longitudinal study of an inter-university first year engineering initiative called 'Connections' that the graduation rates five years later for both males and females improved for two consecutive intakes: 1994 and 1995. The improvement for the females in the 1995 cohort was quite dramatic: 90 per cent retention compared with 64 per cent before the innovation.

An evaluation of the US Gateway Engineering Education Coalition's (eight engineering schools) first year curricular innovations also reported improved retention rates of both female students and all students (Fromm 2003; Dym et al. 2005). The new programs '. . . integrate communications skills, teaming and interpersonal skills, and the ethical dilemmas faced by engineers . . . and bring issues to life in real context . . . ' (Fromm 2003: 117). In 1999, the retention rate for female students was 90 per cent compared with 68 per cent in a national sample, and for all students the corresponding figures were 86 per cent and 70 per cent (Fromm 2003). The Gateway Coalition's curriculum initiatives are examined in more detail in Chapter 7.

The evidence available therefore confirms that courses with recognisably inclusive features have succeeded in improving the learning experience for female students, and their retention, progression and success rates. However, this is an under-researched area and engineering educators who trial or adopt the inclusive practices recommended in this book are urged to evaluate their innovations and to disseminate their findings to other practitioners.

SUMMARY

This chapter has provided some ways of thinking about teaching and learning and demonstrated the importance of considerations of identity, self efficacy and learning styles in achieving effective learning and teaching outcomes. It has focussed on developing the understanding that we all learn more, and more effectively, by being stretched and challenged; and the different perspectives, objectives and experiences which can be contributed to a class by students from diverse backgrounds can contribute to this process. We have defined what we mean by gender inclusive curriculum and considered the theoretical understandings behind it as well as the research that demonstrates its effectiveness in science and engineering education. We have explained that making the engineering curriculum more gender inclusive should not be regarded as a process of remediation but rather as a means whereby the latest advances in understanding the processes of teaching and learning are brought to bear on this particular area in ways that will resonate with the socio-cultural context of the current student body.

Unfortunately the current situation is that there has been little change in most institutions in the way that the gender dimension has affected the tertiary curriculum since a 1998 project concluded that:

> In male-dominated disciplines the curriculum as the 'delivery of traditional facts to students in traditional ways' remains the dominant model. In other disciplines, where women are present in greater numbers, gender dimensions have been considered, seen to add depth to academic pursuits, and been well received by students.
>
> (Bagilhole and Goode 1998: 456)

We believe that we can do better in engineering education by embracing a gender inclusive curriculum. Most of the research covered here has concentrated on student responses to engineering education. In the following chapters we turn our attention to the faculty roles of designing and teaching the courses, preparing the materials, organizing the classes and engaging with students in the laboratories and classrooms. Chapters 5 and 6 will directly address these issues and constitute an answer to the ongoing question this topic provokes: *how* do we make the engineering curriculum more inclusive?

5 Design and Implementation Strategies for Gender Inclusive Engineering Curriculum

INTRODUCTION

Given the current situation for women in engineering outlined in the previous chapters, the case for curriculum adaptation is compelling, and we have already pointed to some developments in gender inclusive education in engineering that are being currently adopted. Chapter 4 has outlined the theoretical basis of the case for gender inclusive curriculum in engineering education and has provided a definition with related components and understandings. We turn now to the practical applications of this theory: the curriculum itself. If you are currently teaching courses in engineering programs to engineering students, or will be in the future, then this chapter is for you! It contains some background information about designing and implementing gender inclusive courses and programs, followed by a wealth of practical advice and suggestions for how to go about it. Examples of course design and implementation are used throughout the chapter, and then in Chapter 6 we provide several more specific course examples. We hope that these two chapters in particular will enable all faculty members to see that they can do something positive with regard to gender inclusive curriculum in their courses and programs, even if it may only seem a small change and even if other members of their departments may not yet be convinced of the value of it. Our view is that anything that can improve the retention and success of female students in engineering is worth doing. There is also nothing to lose by trying, since, as we showed in Chapter 4, an inclusive curriculum improves the performance of all students, not just females.

DESIGNING AND IMPLEMENTING A GENDER INCLUSIVE CURRICULUM

Designing a gender inclusive curriculum usually involves considering all of the components of the curriculum listed here, in approximately this order:

- the assumptions made about the perspectives, experiences, values and backgrounds of the students
- the aims and objectives of the program or course
- the forms of assessment
- the content
- the teaching and learning methods
- teaching practices
- the learning environment, including the resources needed.

Making any of its components more gender inclusive can help to make a course more gender inclusive, but ideally all of its curriculum components should be incorporated into this process. Making a program more gender inclusive requires consideration of its overall assumptions, aims, objectives, and students' learning experiences, in addition to the inclusivity of its individual courses. Though a program can initially be made more gender inclusive by adding gender inclusive courses or strands of courses to an otherwise 'traditional' curriculum, this approach should be regarded as only a step along the way towards a more genuinely inclusive program. In a genuinely gender inclusive program, every course is gender inclusive, within an overall gender inclusive program design.

A considerable part, but not all, of the curriculum as experienced by a student is planned in advance of any student enrolment. This is usually a formal university requirement relating to the process of program and course development and approval. The outcome of this process can be described as the program (or course) design. All of the curriculum components listed earlier can be considered to be part of the design process, but the primary focus in the design stage will usually be on the assumptions, aims and objectives as well as broad determinations of content, assessment, teaching and learning arrangements and resources required for each course.

Once the teaching process starts, the curriculum is continually modified and adapted by the faculty members actually teaching it, in response to the abilities, interests, needs and concerns of the students, the teachers and other stakeholders. The iterative nature of this process can be likened to that of engineering design. All of these influences on, and modifications to, the curriculum which take place after the formal program or course design stage may be described as the implementation phase. This phase will determine in greater detail the content, the forms of assessment, the teaching and learning methods and the resources required. In addition, there are always some unstated, even unconscious, elements of the learning environment which do not appear at all in the formal program or course design but nevertheless can significantly affect student learning experience—this is sometimes referred to as the hidden curriculum. The learning environment includes features such as the physical environment (the room and its equipment) and the faculty-student and student-student interactions, which will incorporate the prevailing attitudes, values and assumptions of the people involved.

The implementation phase is critical because, however gender inclusively the curriculum may have been designed, it will not be experienced as inclusive by the students unless the teaching and assessment methods are implemented along inclusive principles and the learning environment is managed in such a way that it is inclusive of all students. If the teacher illustrates the applications of the theory, for example, in an entirely western, male context (the use of motor vehicle examples in engineering mechanics is a typical case) the course will not be gender inclusive. Likewise, if a particular group of students is allowed to dominate the class or tutorial, the course is not gender inclusive.

Since there is considerable overlap between the design and implementation phases of a curriculum, ideally the person who will eventually teach a course should always be involved in the course design process. The tables that follow provide some suggestions and examples to think about during the design process.

In many cases the ideal situation does not occur and you may only be involved in implementing a curriculum that you have not designed. If this is the case and you have been assigned to teach a course that has been designed with gender inclusivity in mind, what can you do in the implementation of that course to ensure that it remains gender inclusive? Conversely, is there anything you can do if the curriculum has been designed without considering gender inclusive issues and you have been assigned to teach it? You have probably been handed a short description of what you should teach, with some specifications about how you should teach, what resources are regarded as essential and what forms of assessment you should use. This will form part, but only part, of the total curriculum. Typically you have (or can negotiate) some flexibility about sequencing the content, adding supplementary examples and applications, and managing the classroom or laboratory. You may decide to negotiate with students over some of these matters. While the assessment structure may be mandatory, exactly what is assessed and exactly how it is done is, again, usually your decision. The discussion and tables which follow also suggest ways in which you can be gender inclusive with all of the additional elements which enter a curriculum when it is delivered or implemented.

SUGGESTIONS AND EXAMPLES OF GENDER INCLUSIVE CURRICULUM DESIGN AND IMPLEMENTATION

In the following sections we consider each of the main curriculum components in turn and provide some suggestions and examples of ways in which you can make these gender inclusive. We have based these 'think about' points, suggestions and examples on our combined many years of experience in course design and implementation in this area, as well as our involvement with the provision of learning support for students who are struggling with courses, sometimes because they feel that the course does not 'include' them. We have included suggestions and examples provided to us by many colleagues from our own and other institutions who also believe in the principles of gender inclusivity. Please pick and choose from among these suggestions and use them as seeds for developing your own ideas in the courses you design and teach. In many cases you may think that the gender component within these examples is not clear or explicit, and that may be true. However, the suggestions are based on theory (as described in Chapter 4), as well as our own and others' experience of curriculum examples that have gender inclusive outcomes. As we have reiterated several times, curriculum measures that are gender inclusive are usually also perceived or experienced by other minority students as inclusive and can improve the learning experience and outcomes for all engineering students.

Within the text we have also incorporated specific examples from two particular courses at the University of South Australia, one of which was involved in the original inclusive curriculum project at that institution (described in Chapter 7) and the other which has been developed since. At least one of us has been involved in these courses with the curriculum design or implementation or both. The courses described have been selected because they demonstrate inclusivity in some, but not necessarily all, of their curricular components. Whether a curriculum is inclusive because of the manner of its implementation, or whether it has been initially designed to be inclusive, faculty members adapt what, and how, they teach on a day-to-day basis. The extent and features of inclusivity depend upon the decisions made by faculty members about what is appropriate for their classes. Making engineering more inclusive is a process rather than a once-and for-all event, and some of the courses from which we have drawn examples may still be in an intermediate stage of this process. In Chapter 6 we have provided several more examples of gender inclusive courses or gender inclusive components of courses.

Assumptions About Students

When designing a curriculum, the first thing to think about is the needs of the group of students who will enrol in this course or program. Why are they taking this program or course, what do they need from it, what are their backgrounds, how much do they already know and what skills have they got? The greater the diversity of the student group, the greater will be the diversity of their backgrounds, needs, perspectives and ambitions. While some program designers may resolve, or overlook, the challenges presented by these circumstances by designing the curriculum to suit the dominant social and cultural group in the class, we argue that this is not only inequitable, it is also inefficient and counter-productive. The program or course should be designed to give all the students who have been admitted to the class a fair chance of succeeding. In addition, the different perspectives, attitudes and values brought by the minority students can be incorporated into the curriculum to extend, challenge and ultimately improve the thinking of the dominant group.

A good starting point is to consider why the students have selected this program or course. Not all engineering students join the program because they have an overwhelming ambition to become engineers.

> Men have the advantage of a gender role that makes engineering a more obvious career choice. The young tinkerer is likely to become aware of engineering as if by osmosis, as he explores the world of mechanics and technology. It is unlikely that his non-tinkering sister will do the same.
>
> (McIlwee and Robinson 1992: 35)

Many students have little idea of what being an engineer involves. They may have been persuaded to study engineering because they were good at science and mathematics at school and in choosing engineering are consciously taking a pioneering, adventurous step into the unknown. Apart from the important implication that the engineering curriculum should include some information about engineering careers (and other careers open to numerate graduates), it is worth considering whether the differences in the backgrounds of students selecting engineering study imply differences in their personalities and hence their learning styles, as discussed in Chapter 4. If so, these students may be motivated by different factors, and to retain the interest of these students the curriculum should acknowledge these differences.

Another assumption to consider is the educational backgrounds of the students. In some universities, the majority of undergraduate students are likely to have come directly from secondary schools. Since secondary schools vary widely in their approach and standards, even in those countries where there are national curricula, there will be a range of educational backgrounds amongst the students, and more so where curricula are determined at a state or local government level. In other universities the class may contain significant numbers of mature-aged students returning to study after several years in the workforce, possibly at a trade level; others who have followed alternative entry pathways from community or technical colleges; as well as international students. The more diverse the educational backgrounds of the students are, the more complex the issue of assumptions becomes.

A criticism that can be levelled at many engineering courses and programs is the assumptions that are made about prior knowledge of skills which are usually regarded to be outside of school curricula, but which are essential for study within the course or program. Examples include the assumption that all students entering engineering are already computer literate, that they are already familiar with mechanical or electronic components such as a car engine or circuit boards and that they are also familiar with the tools needed to work with these components (commonly called tinkering skills).

We need to consider whether the assumptions that we are making about students' prior knowledge are valid even for the majority of the students in the group. For example, male students may be far less likely to have tinkering experience today than they had 20 years ago, since they are now more likely to spend time playing with Nintendo and computer games than building meccano constructions or crystal radio sets. Even if the assumptions we are making are valid for the majority of the students, we still need to consider how to enable students with perceived gaps in their knowledge and skills to catch up with the majority of students in the group, without singling these students out as needing remedial help.

Table 5.1 summarizes some ideas and suggestions for you to consider when designing your program and courses, with respect to the assumptions you make about your students and the ways in which you can make your teaching more inclusive in this regard.

Example of Assumptions from the Course **Sustainable Engineering Practice (SEP)**

This course has been designed from the outset according to the principles of gender inclusivity. It is a compulsory course for all first year engineering students at the University of South Australia and was first taught in 2008, although it built on a previous course in the civil engineering program that had been taught for several years by one of the authors. It provides an introduction to the engineering profession and addresses social

Table 5.1 Suggestions for Designing a Gender Inclusive Curriculum—Assumptions

Assumptions
Think about
• whether students from diverse backgrounds have the same experiences and interests as 'traditional' engineering students • whether all (or any) have tinkering experience • students' previous access to computers and their levels of computer literacy
Suggestions
• include introductory 'how-to-use' laboratory and computer sessions as an integral part of the program for students who have had limited access to computing facilities, or who have never had the sort of experiences often assumed, like playing with mechanical or electronic toys, or dismantling a car engine. • ensure that these sessions are open to all students rather than being seen as remedial sessions for some students only.

and environmental concerns. It acknowledges the contributions made to engineering and technology in other cultures. SEP develops students' communication and collaborative skills as well as technical and management skills, and it employs a variety of teaching, learning and assessment practices.

The student body in engineering at the University of South Australia is very diverse, coming from a wide range of socio-economic, language, nationality, school and work experience backgrounds. The engineering disciplines available are civil, electrical and electronic and mechanical/manufacturing, and the overall percentage of women in engineering has remained around 11 to 12 per cent in recent years. Due to the wide range of backgrounds of students, the course designers endeavoured to make very few specific assumptions about the students taking the course and to provide as many additional support resources as possible to assist students.

When the course was first delivered we assumed that many students would not be familiar or comfortable with skills such as writing in an academic environment (for example avoiding plagiarism, correct referencing, use of headings) or working in groups, and that they would not understand the importance of communication skills in an engineering career or be comfortable with oral presentations. Consequently we provided lectures and extensive resource material on these topics, even though we also knew that some students from particular secondary schools probably were familiar with all of this. This produced mixed responses, as illustrated by these divergent comments in some students' reflective journals:

> In SEP over the past few weeks I've found it similar to subjects I've done in Years 10 and 11 (Work Education and Careers & Lifestyles) where we've gone through resumes, different personal qualities, group work, etc. The only difference is that SEP is focused on Engineering. The lectures have basically been pretty much the same as what I've already covered.
>
> (an Australian student, attended local secondary school)

> The SEP classes were really slow paced in the first week which I thought was good considering all this new information had been dumped on us. The second week was a hard hitting week. With a huge work load from every course it was defiantly (*sic*) make or break from the start. As far as the SEP classes went we were still going through the basics like making sure we use good speeling (*sic*) and grammer (*sic*). The real highlight for the course so far was the career skills we will learn, great for the resume.
>
> (an Australian student, attended local secondary school)

> Sustainable Engineering Practice was a great experience for me. Researching, understanding, writing reports, writing a Résumé and looking for the areas I need improvements at and try to fix them . . . I had one of my best times studying this course. Most important thing is that I learnt many things during this semester. I am afraid that there will be no chance to write a reflective journal again for this course so I have to say that group assignment was very useful for my personal good and for my future career good. It helped me to get over my weaknesses like talking in front of a group of people and working in group, these two things was like a nightmare at the beginning of this semester but now, I AM ALRIGHT. I would like to take this opportunity to thank our lecturer, tutor, course coordinator and everyone who participate to make this course such a successful and useful for all first year students. THANK YOU VERY MUCH.
>
> (an international student)

These quotes illustrate that the consequences of NOT making assumptions can range from boredom and frustration for some students, to the provision of useful to very important learning experiences for others.

We were able to assume that students would have ready access to computers. For those without computers at home, the university has 24-hour computer pools available on each campus and other pool rooms that are open during normal university hours. For those students not so familiar with online environments, there are orientation sessions conducted before the commencement of the course which are available and encouraged for all students. Part of the assessment within the course is a reflective journal written as a blog and the development of a wiki by each student. We provide detailed online instructions and examples for students to use to develop their blog and wiki pages, but we found that we had over-estimated some students' ability and familiarity with computers and several students had difficulty and required additional help to get started.

In response to the student feedback and our evaluation of the issues that arose during the course as a result of our assumptions, we made changes to course delivery in 2009. Some of the more basic material included in lectures was moved to online access or incorporated in tutorials, but most was retained. The hands-on project work was introduced earlier and the other content was more fully integrated within that to provide more connections and relevance for those who were bored. For the computer familiarity issue we provided compulsory help sessions in computer pools at key times in the semester. All students were required to attend at some time during the week, but if they were not having problems they could just sign off and leave. Those requiring more help could stay as long as they needed.

Aims and Objectives

The aims and objectives of the program or course are the most important part of program design. Until we have clarified what we want our students to learn and how we will require them to demonstrate what they have learned, we cannot begin to develop the content of the program or select the teaching and learning strategies we will employ. Another purpose of expressing the aims and objectives is for the students' benefit, so that they know what will be expected of them.

> The purpose of expressing aims and objectives is to improve the quality of education, in two senses. The activity should enable teachers to think more critically and deliberately about student progress, and the manner of its connection with what they do in their teaching. Second, the results of the exercise should make clear to students exactly what they have to learn to succeed, and what they can leave aside.
>
> (Ramsden 2003: 125)

Yet another purpose of explicitly stating the aims and objectives is to provide a framework for students' evaluation of a program or course. The aims and objectives of an inclusive curriculum will incorporate in some way the consideration of social and cultural influences on the construction of knowledge and on scholastic and professional procedures and practice. Table 5.2 offers some suggestions and examples to consider when designing the aims and objectives of a gender inclusive engineering curriculum.

Table 5.2 Suggestions for Designing a Gender Inclusive Curriculum—Aims and Objectives

Aims and Objectives
Think about
• how to integrate technical understanding with society's needs • developing students' awareness of international, multicultural, gender, indigenous and other perspectives in engineering and technology • preparing students for professional practice in a multicultural society and global economy.
Suggestions
• include social, environmental and global aims and objectives with the technological and professional ones.

Example of Aims and Objectives from Sustainable Engineering Practice

Aim

This course aims to introduce students to the profession of engineering through course and project work and interaction with industry that will help students to develop some of the essential skills needed for sustainable professional practice.

Objectives

On completion of this course, the student should be able to

- Demonstrate a broad awareness and understanding of the various discipline areas of engineering and the role of the engineer in modern society.
- Appreciate the social, political, environmental and economic context in which engineering is practiced.
- Identify important attributes of engineering graduates and use them to evaluate their own personal strengths and weaknesses.
- Prepare a written report and give a short presentation on the outcomes of a basic engineering project.
- Identify the characteristics of effective teams in an interdisciplinary environment and demonstrate their ability to successfully contribute to a team and to evaluate their own and other team member contributions.
- Locate, evaluate and use information from a range of sources to develop a coherent argument which is correctly referenced.
- Demonstrate the application of ethical standards and sustainability principles to the solution of engineering problems.

Assessment

We suggest that course and program designers should ask themselves, immediately after setting the aims and objectives, how they will know that the objectives have been achieved (Brown et al. 1994). Clearly, the answer is by assessing the students, and therefore the structure of the assessment of the program or course should be the next step in the process of curriculum design and development. Only after determining the assessment structure is it justifiable to decide the detail of the content, and the teaching and learning methods.

Murphy (1993) argues that there are two equity aims in assessment, namely to ensure that the assessment practice enables individuals and groups to reveal their achievements and that interpretations of results are valid and just for all groups. A number of implications for equitable assessment arise from the previous discussions about the aims and objectives of a curriculum and the discussion of variations in students' preferred learning styles in Chapter 4. We believe that the structure of assessment should clearly relate to the program or course objectives. If, for example, one of the objectives is to appreciate the benefits and challenges of working in a group to tackle problems in engineering, then students' appreciation of these benefits and challenges must be assessed in some appropriate way. Also, if students have been provided with a variety of learning experiences in recognition of the multiple perspectives contributed to the class by their diversity, the assessment process must reflect this. Yet another factor to be considered is the range of students' preferred learning styles. For example, some students may be more proficient speakers than writers (although some will be proficient at both), and since speaking skills will be highly valued in almost every branch of professional life, we suggest all students should be assessed on speaking skills, thus providing those who speak well, as well as those who write well, with opportunities to excel in the assessment process.

Gender differences in students' performance in science and engineering assessment exercises have been ascribed to the contexts in which questions and problems are set and the assessment methods used. Many women have difficulty with questions which assume a masculine experience acquired outside class (familiarity with the mechanical parts of cars or with male sports), or where the relevance to their own life experience is obscure.

> It has been demonstrated over and over that women do not do as well solving maths problems based on a male sport simply because they have to work out first what the question is about in a way that the men do not.
>
> (Nightingale and Sohler 1994: 31)

Armstrong recommends that exam questions should include some set in the contexts known to be of interest to women: with social and environmental implications for example. She also finds evidence that females have more difficulty than males in understanding mathematical-type problems expressed only in words, without a diagram (Armstrong 1995, citing Johnson 1984). Regarding methods of assessment, Murphy finds that 'There is a considerable body of evidence that multiple-choice items advantage males', and 'evidence of this nature should be considered when selecting types of assessment tools' (Murphy 1993: 281). This was also noted by Gilligan (1982) who found that girls usually wanted to opt for 'it depends' rather than make a choice, and that girls tend to outperform boys in written exercises that allow for less tightly defined solutions as well as assessments that allow girls to use their more developed verbal skills to good effect.

We also recommend that time is allocated during a course to explaining the assessment procedures in detail, making it quite clear what is required for a pass and higher grades. In addition, clear and consistent assessment pro-formas for assignments that are distributed to students at the commencement of a course and then used to provide feedback for the assessment are an excellent idea. These forms detail the components of the assessment, the proportion of total marks allocated to each component and provide an indication of the level of work within each component that would be expected for each grade level. They are very useful for both students and faculty members in ensuring clear and consistent assessment procedures for all. Some food for thought and suggestions about assessment in your design of a gender inclusive curriculum are included in Table 5.3.

Table 5.3 Suggestions for Designing a Gender Inclusive Curriculum—Assessment

Assessment
Think about
• the evidence that some assessment methods favour some social and cultural groups • the need for assessment modes to reflect the teaching and learning methods employed • matching assessment techniques to the different teaching and learning arrangements which develop particular attributes of engineering graduates • whether the context of questions or assignments may favour some groups of students over others • the fact that individual or cultural differences, or a disability, will result in students having different skills and therefore different ways of excelling.
Suggestions
• provide opportunities for students to demonstrate their learning in a variety of ways, including modes which tend to be encouraged and developed in other cultural environments • encourage students to think about their learning and to select among assessment strategies in order to optimize their results • use a mixed portfolio of assessment methods • don't overload students with too much assessment • consider the relevance and the appropriateness of the context of your assignments and exam questions • sometimes allow students to choose different ways to present their assignments, for example orally or written.

Example from the Course Sustainable Engineering Practice

This course employs a variety of assessment strategies that allow for early intervention and interaction with students. Assessment consists of:

- An individual report on the role of engineers (15 per cent)
- An individual online portfolio (45 per cent)
- A group report and oral presentation on the major project (40 per cent).

The aim of the individual report from the student viewpoint is to gain an understanding of the role of a professional engineer and the value of engineering ethics, as well as to further develop communication skills. From the inclusive course design viewpoint, the aim is to provide an opportunity very early in the students' university experience to gauge their competence with written work. All students assessed to be at risk in this regard after this assessment is marked are provided with access to additional learning resources and advised to attend learning support sessions to develop their writing skills.

The individual portfolio involves the development of a wiki (website) and a reflective blog. The wiki components include critiques of online news articles on engineering and sustainability collected by the student, summaries of online job advertisements for graduate engineers addressing the personal attributes being sought, a resumé, career plan and ongoing professional development plan developed by the student and reports on the industry speaker sessions within the course. The blog is required to have at least one entry per fortnight reflecting on what the student has learned from the course and how they have contributed to the group project. The blog and wiki are private, with only the student's tutor invited to have access, and the tutors are required to provide feedback online at regular intervals.

The interdisciplinary group's written report and oral presentation relate to the major project for the course which is described in the next section. Peer assessment is used to moderate the marks of individuals within the group where justified.

Example from the Course Electricity and Electronics

This course was taken by first year electrical and mechanical engineering and computer science students at the University of South Australia for many years, although it has undergone some name changes and some further modifications when the common first year was reintroduced in 2008. The course team undertook a major review of the course as part of the university's inclusive curriculum project (described in Chapter 7) and subsequently implemented several modifications to improve aspects of inclusivity. A range of assessment methods were employed, including tutorial quizzes in the first

10 minutes of each tutorial, team reports of practicals, a multi-choice test at mid-semester and a final examination. Students were also able to earn a 5 per cent bonus by contributing to the development of the course by offering new tutorial questions or laboratory exercises, for example. The format of the final exam was an example of an inclusive approach to assessment, as it was designed to acknowledge student diversity and a range of learning styles and thus differed significantly from the typical electrical engineering exam which focuses on circuit solutions (Ayre and Nafalski 2000). The exam was a mixed-mode, two-hour closed-book examination. It contained essay questions which required a prior literature search, giving students a chance to demonstrate both their literary skills and technical knowledge. There were also graphical questions using schematics and oscilloscope screen images, and conceptual questions as well as a limited number of circuit solutions. Students were given advance notice of 18 topics to be examined and at the exam itself they selected 10 out of the 18 questions. Interestingly, the course team's own research on students' selections confirmed the existence of a gender bias in question selection, with males tending to prefer the rigorous circuit solutions (Ayre and Nafalski 2000).

Content

The content of an engineering course (or syllabus) typically comprises some standard applications as well as the underlying theory, for example the basic laws of electricity as well as their applications to the design of transistors or a stereo radio amplifier. In addition, in the teaching process, the faculty member will usually expand on the syllabus by introducing further applications and examples. In the disciplines of science, mathematics and engineering, the curriculum content can therefore be thought of as comprising two components: the underlying theory and some standard applications (sometimes termed the 'primary content') and other applications and examples (secondary content) (Warren 1989). Determining the curriculum content is part of both the design and the implementation of a curriculum, but frequently the primary content is determined as part of the design process and the secondary content as part of the implementation. However, we would point out that during curriculum development, faculty members may find it helpful or easier to first consider how to make the secondary content more inclusive (by thinking about examples and applications that are gender inclusive), then consider the implications of these examples for the primary content.

Many engineers and scientists maintain that their core curriculum (or primary content) is based on universal laws and is not therefore subject to cultural or gender bias, but as Pawley (2004:7) notes, 'To make our classrooms truly gender-sensitive, the content cannot remain sacrosanct'. Rosser, for example, points to alternative views which challenge belief in gender neutral content:

> Some scientists, influenced by scholarship in women's studies, the philosophy and history of science, and psychology, have begun to recognise that gender may influence science. Kuhn and his followers suggested that all scientific theories are the products of individuals living in a particular historical and social milieu.
>
> (Rosser 1995: 5)

> In the teaching of science, most instructors underline the importance of the scientist's objectivity in approaching the subject of study. This is thought to be necessary to establish scientific rigor . . . Feminist critics . . . as well as practising scientists . . . have pointed out that the portrayal of the scientist as distant from the object of study masks the creative, interactive relationship many scientists have with their experimental subjects.
>
> (Rosser 1995: 7)

In contrast to the traditional picture of the objective impersonal scientist, Fox Keller (1983: 179) described the preferred position of Nobel Prize winner Barbara McClintock as 'letting the material tell you' and 'developing a feeling for the organism' as perhaps more typical of women's approach to doing science. Harding adds:

> If . . . science is presented as impersonal, detached (objective), abstract, law bound and enabling control, then it, especially, matches the needs of immature or emotionally reticent males and enables these, in larger numbers than from other groups, to progress through its study.
>
> (Harding 1996: 116)

Roberts and Lewis describe two different approaches to the construction of knowledge and ways of knowing in engineering and science:

> One approach is more like the traditionally known, objective, rule-seeking ways of evaluating, proving or disproving truth. The second approach seeks understanding and meaning from the individual's perspective—connected procedural knowledge—for people who look for connections between events and personal factors such as circumstances and background experiences.
>
> (Roberts and Lewis 1996: 159)

This distinction between approaches maps neatly onto Gilligan's description of the outcomes of typical gendered socialization that takes shape in thinking patterns with the boys looking for the abstract rules and the girls seeking to remain connected to others in their context (Gilligan 1982).

The secondary content, namely additional applications and examples, is a critically important area that can have a significant influence on the way in which an engineering curriculum is experienced by women. If everything else in the curriculum design has considered gender inclusive principles but the faculty member who teaches the course uses exclusively male-focussed examples and applications, then it is likely that much of the rest of the design will be a wasted effort. Conversely even if nothing but the secondary content has considered gender-inclusive principles, a wise choice of examples and applications can still make a huge difference to female students' perceptions and experience of the course.

Table 5.4 provides some suggestions and things to think about when designing the content of a gender inclusive curriculum.

Example from Sustainable Engineering Practice

Many more examples of gender inclusive content will be presented in Chapter 6, but one is worth mentioning here. Sustainable Engineering Practice (SEP) utilizes content that has now been incorporated into the majority of

Table 5.4 Suggestions for Designing a Gender Inclusive Curriculum—Content

Content
Think about
• incorporating the interests and experiences of diverse social and cultural groups • challenging a uniform view of knowledge • developing, in the classroom and the laboratory, cooperative, communicative, creative and critical skills as well as technical, logical, analytical, and competitive skills • providing students with 'open-ended' opportunities to relate, apply, and generalize from, and hypothesize with, the knowledge and skills they are acquiring • the fact that students who withdraw from engineering programs often complain of lack of creativity and relevance; and of being bored • relating theory to the 'real world' (but pay attention to the issues raised in the following two dot points) • typical examples of the applications of theoretical material (e.g. car engines) which may be unfamiliar or uninteresting to female students • analogies (e.g. sports) that are often gender and culture specific • the traditionally dominant emphasis on the contributions of Western males to the development of technology which can leave female students feeling somewhat alienated and excluded • 'Why have we selected these topics to 'cover' and discuss? Is it because they have been traditionally included, and if so, by whom, for whom, and for what purpose?' (Pawley 2004: 7)

continued

Table 5.4 continued

Suggestions
• include applications of technology in different physical, cultural and social contexts: e.g. electrical/electronic appliances where the power supply is unreliable; and bio-medical, as well as military, applications • discuss the ways in which technology has improved peoples' lives • include reference to alternative scientific methodologies, e.g. feminist science • give students investigative problems for which they need to devise their own experiments, rather than standard laboratory exercises with an expected outcome (or 'right answers') • make creativity and innovation an integral part of the program • include some cross-disciplinary study • focus not only on the technical solutions of engineering problems, but also on the purpose and the social benefits • use examples of general familiarity and interest, e.g. health-related - devices for the disabled, sanitation, diet, domestic applications and musical contexts • use analogies only if they will be understood in the same way by all the students present • include women's contributions to technological development. Some examples are: Florence Nightingale (sanitation reform, statistics), Ada Lovelace (collaborated with Babbage on the design of the analytical engine), Alice Tredwell (constructed the Bhore Ghaut railway in India), Emily Roebling (Brooklyn Bridge construction). See Alic (1986) and Ambrose et al. (1997) for others.

first year engineering programs across Australia. This involves a student design project: the *Engineers Without Borders challenge,* developed and supported each year by the Engineers Without Borders (EWB) organization in Australia (EWB 2009). It is undertaken by student teams of four to six members and requires them to develop conceptual designs for multidisciplinary, international, sustainable development projects being undertaken by the EWB group. Each year the project is focussed on a different community; for example in 2009 the project looked at the Tonle Sap region of Cambodia. Many resources are provided online by the EWB organization and a representative from the organization also introduces the project to the students at the start of the semester. Students may develop their own specific project from a wide range of suggested areas across all engineering disciplines. In SEP the project constitutes 40 per cent of the course assessment and runs throughout the semester. The content of this project easily satisfies the criteria for gender inclusivity, incorporating a large number of the elements in Table 5.4.

Teaching and Learning Methods

Teaching and learning methods comprise the teaching and learning structures and processes we select as appropriate for the courses and the classes we teach. These structures and processes may include lectures, problem-based or project-based learning, small group teaching, individual or group assignments or projects, laboratory practicals, case studies, field trips, peer-assisted learning and more. Discussions about these, and other, teaching and learning methods in higher education may be found in texts such as Ramsden (2003).

The effectiveness of teaching and learning methods may be moderated or enhanced by the personality characteristics, and expectations of the teaching and learning experience, of both the teachers and the learners. Our teaching methods are merely the instruments we select to implement our own beliefs about the role of teaching and how to teach effectively, which may be termed our teaching styles. These teaching styles, like the methods with which we implement them, have been formed from a combination of influences, for example our personalities, intuition, our own experience of teaching and being taught, cultural norms, as well as what we have learned about teaching and learning, formally or informally. All of these attributes will have been influenced to a greater or lesser extent by cultural and social factors. By a similar process, students develop individual identities and learning styles, influenced by individual and cultural norms, as discussed in Chapter 4.

Since texts on teaching methodologies do not often address the issues raised by cultural influences on teaching and learning styles, we provide a brief discussion here. The purpose is to suggest that engineering faculty members should consider how social and cultural differences in teaching and learning styles may affect students' performance and success. It is therefore worth considering not only how to counteract any disadvantage to certain groups such as women, but better still, how to use these differences to extend the experience of all students.

Harding (1996) draws attention to a study undertaken in the 1970s (Galton 1981) which explored gender differences in teaching styles in science lessons in schools, which may have implications in tertiary education. Galton identified three teaching styles: problem solvers, informers and enquirers. The problem-solving style challenges students to generate hypotheses and solve problems. This style was used more by male teachers than female teachers but was the least popular of the three with female students. The author suggests this unpopularity arose from girls' tendencies to approach problems from alternative perspectives, while the teacher expected a single correct answer. Most of the women teachers in the study used the pupil-centred enquiry style. It was found that the informer style (focussing on recall and application of facts) and the pupil-centred enquiry style were equally popular with the female students. While subsequent research (Severeins and Ten Dam 1994) has failed to find consistent differences in teaching approaches between males and females, the important feature to emerge from this work is a heightened awareness on the part of the teacher and the learner of the degree to which gender can enter into the process and either promote or hinder effective teaching and learning.

Table 5.5 provides some ideas and suggestions about teaching and learning methods that should be considered in designing a gender inclusive engineering curriculum.

Example from the Course Electricity and Electronics

Teaching and learning methods used in this course included traditional lectures, tutorials and small group practicals, supplemented by multimedia such as online video, computer software packages and electronic communication via bulletin boards and discussion groups. Problem-based learning was used in the digital component of the course with students given the task of designing and then demonstrating a working prototype of a digital logic circuit to control a pump filling a water tank. Supplemental Instruction (SI) was also available for students in *Electricity and Electronics*. SI is a semi-formalized system developed by Deanna Martin at the University of Missouri (Martin

Table 5.5 Suggestions for Designing a Gender Inclusive Curriculum—Teaching and Learning Methods

Teaching and Learning Methods
Think about
• the fact that active learning strategies are generally regarded as leading to more effective learning than passive methods like lecturing • using a variety of teaching methods to accommodate a range of learning styles and develop a range of skills in all students.
Suggestions
• aim to extend the preferred learning styles of all students by employing a variety of teaching methods • teaching and learning arrangements could include, for example: problem-based learning, mini-projects, peer-assisted learning, case-studies, computer-based learning • make sure there are opportunities for collaborative learning, as well as learning in competition with others • give open-ended laboratory investigations and computer simulations (to illustrate, for example, invisible phenomena such as magnetic fields) as well as exercises with 'closed' expected results.

and Arendale 1994), in which students form their own support groups to integrate what-to-learn with how-to-learn. Groups address, for example, note taking, questioning and revision techniques. The groups were led by specially trained, high-performing students from previous years. Despite the name, SI was not instituted in the course as a remedial system; any student was able to join, and the principal aim was that students should become autonomous, reflective learners able to work in teams (Ayre and Nafalski 2000).

Teaching Practices

As students proceed through any one educational system, they become increasingly socialized into the teaching and learning methods of that system but face particular challenges when transferring to a new system. The teaching practices at university or college level will usually differ substantially from those at secondary or high school level. So not only do all students have to make significant adjustments to their approaches to learning during their first year at university (a time often identified as critical for students' retention and eventual success), they must also adapt to new styles of teaching. Even greater challenges are faced by students from backgrounds in which a different culture operated in their prior studies, whether that be gendered or otherwise. It is therefore important, particularly when teaching first years, to be aware that students from diverse backgrounds are likely to approach their learning in different ways. These ways have been formed by personality characteristics, as well as students' cultural backgrounds, including the ways that they have been encouraged to learn in their previous educational institutions. It helps also to have some knowledge of what may constitute these different learning styles, and we have discussed this in some detail in Chapter 4. The following statement by Felder and Silverman (1988) summarizes the unhappy situation in many engineering departments:

> Learning styles of most engineering students and teaching styles of most engineering professors are incompatible in several dimensions. Many or most engineering students are visual, sensing, inductive and active and some of the most creative students are global; most engineering education is auditory, abstract (intuitive), deductive, passive, and sequential. These mismatches lead to poor student performance, professorial frustration, and a loss to society of many potentially excellent engineers.
>
> (Felder and Silverman 1988: 680)

We are not of course suggesting that faculty members should try to adapt their teaching to suit each individual's learning style, merely that they should develop an awareness of the range of different styles and provide all students with opportunities to employ, and extend, their own preferred learning styles. Some ideas and suggestions to assist you to do this are provided in Table 5.6.

Table 5.6 Suggestions for Implementing a Gender Inclusive Curriculum—Teaching Practices

Teaching Practices
Think about
• the fact that women tend to start engineering studies with a lack of confidence about whether they have had adequate preparation • evidence that this lack of confidence amongst women not only persists but even increases during a program • encouraging students to take more responsibility for their own learning • how to ensure that all students receive equal attention and challenge, without over-compensating • whether it is always beneficial to students to allow them to work in the same-race, same-gender groups they often select for themselves • whether students are experiencing language difficulties • recognizing that some students are more comfortable with cooperative learning styles than competitive ones and vice versa.
Suggestions
• check that technical and jargon terms are understood by everyone in the class • be sensitive to levels of student confidence and give praise and encouragement where appropriate, as well as criticism • set up group activities, and monitor these to encourage all students to take an active part • move around in laboratories, make sure roles rotate, sometimes mixing up groups • use a range of media to provide students with essential (program, and administrative) information: e.g. electronic, as well as written notes. • allow students plenty of time to respond in class—not everyone wants to respond spontaneously; some prefer to compose their response before replying.

The Learning Environment

The reasons that many women feel uncomfortable in science, engineering and technology programs have been well researched (the term 'chilly climate' is often used in this context). At least some of these reasons may well be shared by other groups who are in a minority in these classes. These factors deserve attention because unhappy or uncomfortable students will not achieve as well as they might in a more supportive environment, and they may even withdraw from the program or course. In Chapter 4 we have given examples of the 'chilly climate' experienced by some women in engineering classes. In

Table 5.7 some ideas to think about and suggestions to consider are provided to make your learning environment more gender inclusive.

Table 5.7 Suggestions for Implementing a Gender Inclusive Curriculum—Learning Environment

The Learning Environment

Think about

- teacher-student gender and cultural dynamics in the classroom and laboratory
-
> Society encourages males to be more outspoken, so in a way tutorials and lecturers cater to that, to the sort of learning that males are more receptive to, and if you are quiet you are intimidated by the outspokenness. If there are only one or two females in the classroom then usually it is the male gender who speak.
>
> (student comment quoted in Lintern 1995: 13)

- student-student dynamics
-
> The students constantly mentioned joking, 'paying each other out', and putting each other down as forming a major part of their interactions . . . Some conceded that it sometimes went too far, and that some students could feel hurt or excluded.
>
> (McLean et al. 1997: 145)

- whether disruptive behaviour by a dominant group inhibits the learning of other students
- whether men and women students select different roles in lab sessions. Does this matter?
- the atmosphere of the classroom—is it friendly and interpersonal or business-like or formal and remote?
- do the teachers emerge as personally involved in the learning process?

Suggestions

- encourage all students to participate in class. If some are not doing so, try to find out why not, (for example, students from some non-Anglo cultures view speaking up as inappropriate behaviour), but also accept that some prefer to remain quiet
- ask students from various groups to help you modify learning activities so that they feel more able to participate
- never tolerate racist and sexist remarks, even if only meant jokingly
- develop some 'rules' for classroom and laboratory behaviour, so that each student has opportunities, and encouragement, to participate in the learning activities and no groups dominate discussions or claim the major part of your attention.

USING GRADUATE ATTRIBUTE FRAMEWORKS FOR DESIGNING AND IMPLEMENTING GENDER INCLUSIVE CURRICULUM

Engineering programs in all countries are required to meet some form of accreditation requirement prescribed by a national or international body. In many institutions generic graduate attributes are also used to guide course and program development. Details of some of these accreditation requirements as well as a generic set of *Attributes of Engineering Graduates* are provided in Appendix A. These attributes could also be considered to define the aims and objectives of engineering programs, which, as we stated previously, are essential in order to clarify the content, assessment and the teaching and learning strategies to be used in engineering education. Hence the *Attributes of Engineering Graduates* also provide a useful framework for faculty members to utilize to develop educational strategies for improving inclusivity in curricula, teaching, learning and assessment. In Table 5.8 we consider each of the attributes in turn and for each one provide a number of indicators expressing inclusivity for the attribute as well as generic educational strategies that can be adopted to achieve the attribute. You may not feel that all of these can be applied to engineering directly but think carefully about this before you dismiss any, as most can find application in at least some fields of engineering. The use of these tables will help to develop gender inclusive engineering courses and programs, even though many of the suggestions may not seem gender specific. These tables are a complementary resource to those provided elsewhere in the chapter.

HOW WILL I KNOW IF I HAVE SUCCEEDED?

Evaluation is a critical part of any change process: how can you tell if what you have done is working? In Chapter 7 we discuss the evaluation of engineering programs and initiatives for inclusivity, but in this chapter we are primarily concerned with course evaluation. How can an individual faculty member or course teaching team determine whether they have succeeded in making their course more gender inclusive? There are many ways in which teaching and curriculum changes can be evaluated, and we would again refer the reader to a more detailed teaching text such as Ramsden (2003) for multiple ideas and suggestions. It is true that it may be difficult to determine exactly what components of gender inclusive design are effective in improving the learning experience and retention of female engineering students, or how to measure these effects. It is also true that the positive effect of only one course may not be sufficient to offset the negative effect of others or of other aspects of the engineering education environment. However, sometimes even one course can make a huge difference to individual students, but if no evaluation is undertaken, how will we know? Evaluation results are essential to reinforce the value of an inclusive approach to learning, teaching and assessment and to suggest how to improve it. They also

Table 5.8 Educational Strategies for Improving Inclusivity in Curricula, Teaching, Learning and Assessment Framed Around Graduate Attributes

A graduate from an undergraduate, professional level engineering degree program has a **body of knowledge** of sufficient depth to begin practice in one branch of engineering and the ability to integrate that with knowledge from different branches of engineering and related professions

INDICATORS EXPRESSING INCLUSIVITY	EDUCATIONAL STRATEGIES
• demonstrate an understanding of the potential of the theories and perspectives of non-dominant groups to contribute to the field • apply knowledge to a range of social and cultural contexts and demonstrate how particular ideas within a field of study have privileged some above others • identify the potential of the theories, content and methodological approaches of the field to influence the needs, interests and perspectives of non-dominant groups • recognize the social and historical influences on the development of the field and in particular the effects on specific social and cultural groups • demonstrate an understanding of the way the accepted approaches to, and outcomes of, research have the potential to differentially affect or inform various social and cultural groups.	• consider the process by which various parts of the field have come to be accepted as legitimate knowledge • provide curriculum examples from cutting edges or at the boundaries of the discipline or professional area to illustrate competing schools of thought within a discipline or professional area • provide examples from the history and sociology of the discipline or profession which demonstrate the problematic nature of the way knowledge has been produced (or excluded) and legitimated (or discounted) • provide examples of knowledge which has been accepted at one point in history but which is not currently seen as legitimate and analyse the reasons for this • evaluate the content with respect to its relevance to, and the interests of, the social and cultural diversity of the student group • encourage students to critique the curriculum in respect to its effect on any particular social or cultural groups • question assumptions about prior knowledge and the implications in catering for diversity of student backgrounds • specify resources which are inclusive of under-represented and/or non-dominant groups

continued

Table 5.8 continued

	• incorporate the contribution of under-represented and/or non-dominant groups in the development and practice of the discipline • provide a framework of learning experiences which takes into account, and validates, the interests and experiences of diverse social and cultural groups.
A graduate from an undergraduate, professional level engineering degree program recognizes the need for, and has the ability to undertake, **life-long learning**	
INDICATORS EXPRESSING INCLUSIVITY	EDUCATIONAL STRATEGIES
• understand the social and cultural dimensions of networks of knowledge and be able to recognize their implications in locating, evaluating, managing and using information • understand the partial and relative nature of their own knowledge and its construction in relation to their historical, social and cultural experiences • recognize the potential for enlarging their repertoire of learning styles to include strategies appropriate in a range of cultural or social groups • maintain a concept of self in relation to, and which is informed by, wider social and cultural perspectives • sustain an intellectual approach which embraces the changing social and cultural professional context.	• structure and stage information literacy into the requirements of assignments • incorporate into the details and discussion about specific assessment the relevant issues of locating, evaluating, managing and using information • as a part of the curriculum, engage students in an analysis of their own readiness for learning and preferred ways of learning and provide opportunities for these personal styles to be discussed in a group context • use a range of learning and assessment strategies providing learner choice within the curriculum • give opportunities for students to maintain a positive self concept by providing opportunities for all students to affirm and use their previous experiences within the curriculum • use a range of registers, language styles and vocabularies in different media to model communication practices which include rather than exclude

continued

Table 5.8 continued

	• respect and build on diverse prior learning experiences, values and goals • specify assumed knowledge and skills and make provision for students to access these • respect, value and make provision for different kinds of student participation.
A graduate from an undergraduate, professional level engineering degree program has the ability to identify, formulate, and **solve engineering problems** in a logical, critical and creative way	
INDICATORS EXPRESSING INCLUSIVITY	EDUCATIONAL STRATEGIES
• use analysis and synthesis techniques which are relevant to particular social and cultural contexts • define and analyse researchable questions from a range of social and cultural perspectives and positions of interest • recognize the cultural and social embeddedness of problems with respect to both their conceptualization and solution.	• provide examples of ways of framing and solving problems from other cultures or knowledge paradigms • demonstrate alternative methods of solving problems which arrive at similar solutions • provide a range of problem situations which arise from the range of circumstances experienced by students • encourage different styles of thinking, problem solving and group participation and ask for personal and/or group analysis of the effectiveness of these styles • employ the diversity of the class as an educational resource • provide opportunities for students to demonstrate their learning in a variety of ways, including modes which have been particularly encouraged and developed in other cultural environments.

continued

Table 5.8 continued

A graduate from an undergraduate, professional level engineering degree program can function professionally as both an individual and as a member of a multidisciplinary team (**Autonomy and teamwork**)

INDICATORS EXPRESSING INCLUSIVITY	EDUCATIONAL STRATEGIES
• recognize that self direction may involve fundamental differences in approach for individuals from different groups • be aware of the social and cultural factors in constructing arguments and negotiating with others • work collaboratively in groups which comprise members from varying social and cultural backgrounds • work in teams which comprise members from varying social and cultural backgrounds.	• as a part of involving students in group work, encourage shared responsibility and the recognition and value of differences in opinion and methodology • encourage both formal and informal collaborative work among students, including peer tutoring or mentoring • provide opportunities for students to play different roles in groups and teams • encourage analysis and appraisal of the effectiveness of the group process.

A graduate from an undergraduate, professional level engineering degree program understands their professional, **ethical and social** responsibilities

INDICATORS EXPRESSING INCLUSIVITY	EDUCATIONAL STRATEGIES
• demonstrate a personal ethical position which includes the recognition and valuing of social and cultural diversity • contextualize the social and cultural origins of technologies and demonstrate their differential impact on particular groups in society • demonstrate any links between 'value-free' and 'merit-based' decision making and social inequality • consider the dynamic relationship between construction of privilege/disadvantage and the ability of discipline knowledge to perpetuate/dismantle social inequality	• consider how concepts of social justice might apply in areas of the curriculum • provide opportunities to analyse the impact of new knowledge or discovery in relation to employing such knowledge in socially responsible ways • provide opportunities for students to analyse what it means to be ethical in relation to using knowledge in their discipline or professional area • provide examples from the field of unethical and/or socially irresponsible action

continued

Table 5.8 continued

• recognize the social and cultural issues associated with environmental sustainability • demonstrate the social and cultural dimensions of community responsibility.	• analyse the process by which unethical and/or socially irresponsible actions from within the field have occurred • include within the program a range of perspectives which value international, multicultural, gender and indigenous viewpoints • provide opportunities for students to role-play other views and perspectives • set assignments which provide opportunities to make social, ethical and environmental connections between the program content and the local, global or professional community • recognize and challenge the power of a curriculum to produce and maintain social inequality • use applications which demonstrate the actual and potential social benefits of the discipline • provide contexts in which professional issues are explored for their social, moral, political and economic implications • specify a variety of teaching, learning and assessment practices which are appropriate to the needs of the discipline and the profession and which acknowledge student diversity.
A graduate from an undergraduate, professional level engineering degree program **communicates** effectively in both a professional and a community context	
INDICATORS EXPRESSING INCLUSIVITY	EDUCATIONAL STRATEGIES
• demonstrate an understanding of the links between literacies, and social and cultural experience	• give opportunities to interact with and report to a diverse range of audiences and be able to analyze appropriateness from the viewpoint of an audience

continued

Table 5.8 continued

• demonstrate an understanding of relevant social, cultural and inter-personal communication issues in socially and culturally diverse audiences • value and promote forms of communication from less dominant groups.	• give opportunities to construct and use knowledge according to the varieties of audience that a professional in the field would need to address • provide examples and analysis of communicating complex issues and problems which reflect a variety of perspectives • provide examples and analysis of failures in communicating complex issues and problems • provide opportunities and resources to enable all students to develop the language skills required for the program • pay explicit attention within the class to human relations aspects of communication.

A graduate from an undergraduate, professional level engineering degree program understands the impact of engineering solutions in a **global, economic, environmental, and societal context**

INDICATORS EXPRESSING INCLUSIVITY	EDUCATIONAL STRATEGIES
• appreciate the intersecting and sometimes conflicting value positions of multicultural communities in professional practice • critique international standards and practices within the notions of best practice of a profession • understand the social issues associated with particular cultural and international contexts • demonstrate an awareness of the complexity of decision making with respect to various sub-groups within an international context • demonstrate a global understanding of issues which involve considerations relevant to particular social groups such as women and indigenous people.	• identify examples of the various value positions in multicultural communities and their implications for the field or profession • search out scholarship in the area from other cultures and provide opportunities for analysis of this scholarship against cultural values and standards • compare social practices between cultural groups.

Note: This table has been adapted from tables originally developed by the University of South Australia as part of their university-wide inclusive curriculum project described in Chapter 7.

provide the evidence needed to encourage colleagues to consider making their teaching more gender inclusive.

Course evaluations in an inclusive curriculum need to be constructed around the principles of inclusivity identified in Chapter 1. From this perspective students would be asked to comment on particular aspects of their experience of the course in addition to the standard items regarding the content, timing and so on. A good part of the rationale for inclusive curriculum derives from its capacity to make students and faculty more aware of features of the teaching experience that impact on the capacity to learn. Student responses can be invaluable in alerting faculty members to taken-for-granted assumptions about their course and the way in which it is presented. Course evaluations are one way in which this potential for improvement can be demonstrated, but at the same time the general principles of course evaluation apply. Thus evaluations should not be very long, should lend themselves to simple responses and should be clearly written in unambiguous prose. The results ideally should be widely shared so that both faculty and students can engage in ongoing discussion about the whole teaching and learning enterprise.

Most universities will have some form of standard course evaluation with standard questions, very few of which will relate to inclusivity. However, there is usually potential to add optional questions to these standard evaluations, or alternatively to undertake a small supplementary evaluation to enable faculty members to explore issues of inclusivity in their courses. Some sample questions to evaluate inclusivity that could be used in this regard would include:

- Did you feel that you could approach your instructor/lecturer/tutor to ask about an issue from the course?
- Were there aspects of the course that assumed previous knowledge of the area?
    - o If so, can you identify some of them?
    - o Did you have this knowledge or did you have to work out how to get it?
- Was there a good degree of collaborative work involved in your experience of the course?
    - o Do you like this degree of collaboration?
- Did the course involve the students both individually and as a group?
    - o Would you have preferred more opportunity to work individually?
- Which component of course presentation for example lectures, tutorials or laboratory work did you prefer most?
- If you are an international student (or consider the situation if you were) what aspects of the course might you have found particularly challenging?
    - o Language
    - o Style of faculty-student interaction
    - o Student-student interaction

- If you had a brother/sister/friend thinking about taking this course would you encourage them to do so?
  - o If yes: what features did you rate positively?
  - o If no: what did you especially dislike about the course?
- Would you be willing to discuss your responses with your instructor? If so please give your name at the end of the session.

One example of course evaluation of gender inclusive strategies was undertaken by Kramer-Koehler et al. (1995), who devised a new core curriculum based on inclusive principles discussed in Chapter 4. Before the program commenced, students' existing computing skills and learning styles were assessed. Evaluation of the course was disaggregated to assess its success with male and female students and students from different ethnic and educational backgrounds. Evaluation measures included a course evaluation questionnaire, learning styles assessments and course grades. None of the features that students found to be most inclusive, such as the use of senior student peers as laboratory instructors, could have been immediately observable to the people connected to course delivery, hence illustrating the desirability of conducting carefully designed evaluations, particularly in the interests of inclusivity.

SUMMARY

This chapter has provided a wide range of practical advice and suggestions that can be used by faculty members who wish to design a new course or program to be more gender inclusive, or who wish to modify an existing course that they are currently teaching to achieve the same goal. We have presented the information in two ways. The first breaks down the curriculum into its key components and offers ideas to think about and suggestions to take up for each component. Some examples from courses taught or developed by the authors are also provided. The second approach uses the graduate attributes framework that is being increasingly adopted in university course development and approval processes and again offers a range of ideas and suggestions that faculty members can consider for their own courses. The practical theme is continued in the following chapter, where we provide several more examples of gender inclusive courses or course components and answer some of the most common questions that faculty members have with regard to gender inclusive engineering education.

6 Gender Inclusive Engineering Courses

INTRODUCTION

This chapter is addressed to engineering teaching faculty. It focusses on how to be more gender inclusive in the engineering classroom or laboratory. The major part of the chapter, *Inclusive Curriculum Course Examples*, uses real case studies to illustrate how curriculum features of undergraduate engineering courses of all types can be made more inclusive. Some of the courses described have been developed or enhanced specifically to improve gender inclusivity; others incorporate content and classroom practices which are inclusive because they succeed in creating an environment of shared learning. The next part of the chapter, *Inclusive Curriculum Snippets*, offers more real examples of inclusive engineering teaching, but in abbreviated, or 'snapshot' form. We then provide *Inclusive Curriculum Suggestions Categorized by Attributes of Engineering Graduates*, which may be helpful for faculty in institutions which require curriculum features to relate to graduate attributes, as well as for some engineering accreditation processes, which use a similar framework. We conclude the chapter with answers to some *Frequently Asked Questions* about gender inclusive teaching raised by engineering faculty members. These case studies and examples can be used by teaching faculty to suggest what can be done tomorrow to improve inclusivity to some extent, as well as how to make a course more thoroughly inclusive over a period of time. We hope that the examples and suggestions also illustrate that changing a course to become more gender inclusive does not involve major investments of time or other resources that you may not have, nor does it require additional content to be added. Rather, it requires you to think differently about what you have been doing.

CURRICULUM IMPLEMENTATION

There are two stages of curriculum development: design and implementation. In Chapter 5 we have discussed the process of designing an inclusive engineering program from the outset, which involves specifying the aims and objectives, content (or syllabus), teaching and assessment methods and so on, and then adhering to these specifications for the life of the program. The implementation of a curriculum is the delivery phase when the designed curriculum is taught, learned and assessed. This is the responsibility of the faculty teaching a course. While the faculty members must stay within the broad framework specified, they usually have some discretion about which parts of the content to prioritize and emphasize, and which to omit. They also usually have considerable freedom to select examples and applications to use and to interpret the teaching, learning and assessment specifications quite broadly. Since it is also faculty members who design the assessments and write the examination papers, they play a very big part in shaping the curriculum.

A curriculum which is designed to be inclusive therefore relies heavily on its implementation stage if it is to be experienced as inclusive by the students. A corollary is that a curriculum which is not specifically designed to be inclusive can always be made partially inclusive when being implemented, since faculty teaching the course have many options to vary the course content (for instance in the applications considered), to exercise choice in the teaching and assessment methods and to manage the classroom environment. Thus, it follows that if you are teaching a course designed to be inclusive you need to know how to implement it inclusively. Alternatively, if your course does not have inclusivity built in at the design stage, but you want to make it more inclusive, this chapter illustrates how it can be done, using actual examples. These selections from classroom practice and our own experience are offered in order to stimulate your imagination and to motivate you to experiment. Then, depending on feedback you receive, such as your own sense of effective teaching, observed or measurable improvements in learning, or students' reactions, you can decide whether to retain these elements within the course, or modify, or reject them if they do not result in improved engagement, retention and success for all students. However, we hope that if you reject one of our suggestions as being ineffectual, in view of the evidence we present of the success of these curriculum features in improving the participation of female engineering students, you would return to the literature to re-examine the model, revise your approach and try again, perhaps with one or more of the other strategies. As Jackie Spears, Director for Science Education at Kansas State University and author of the textbook *The Fascination of Physics*, points out:

> If disadvantage arises from the accumulation of relatively small differences, then "undoing" that disadvantage can occur through similarly small changes. This starts with understanding gender schema and how they operate, "seeing" the ways that these schemas contribute to classroom interactions that discourage females and then making small adaptations designed to change the classroom environment.
>
> (Spears 2008: 136–137)

Here Spears refers to the concept of 'gender schema' originated by Sandra Bem (1981) which relates the individual's development of an understanding of gender as a system of classification and the self as a participant within that system and suggests that all learning is interpreted through a gender schema or gender framework. The task for the engineering educator becomes one of ensuring that classroom learning both acknowledges and enlarges the gender schema that students bring to their work. In the following section we make explicit the ways in which this process might begin.

INCLUSIVE CURRICULUM COURSE EXAMPLES

Our examples have been drawn primarily from the international literature, in some cases elaborated by our discussions with the faculty involved, or by attending lectures and presentations by these faculty members. In addition we have included some examples that we have taught ourselves or been involved with developing at our own institution. In making our selection we have not attempted to apply any standard test or measure of gender inclusivity, nor do we imply that the courses used from our own institution are exceptional. Any existing evaluation criteria of inclusivity are necessarily very institution specific, and criteria used at any of our institutions may be inappropriate to apply to courses and programs at other institutions. The purpose of this chapter is to illustrate how the principles and guidelines set out in Chapters 1, 4 and 5 can be adopted or trialled by classroom teachers, adapting them appropriately to their own classes. Readers involved with program and institutional management who are interested in integrating gender inclusivity into the formal processes of course and program approval are referred to Chapter 7 where these matters are addressed.

While some of the courses and curriculum features that follow have been designed or revised with the specific intention of improving their gender inclusivity, others have been selected because they include content or employ teaching and learning strategies, which meet the criteria for gender inclusivity that we have specified earlier in this book, even if there has been no mention by their authors of an intention to address gender issues. In almost all cases, the course or curriculum features to which we draw attention have been evaluated and shown to have improved the engagement and learning of all students, including women. Curriculum features we have looked for include linking content to everyday student experiences, student centredness, active and cooperative learning, developing communication skills and the inclusion in the curriculum of ethics and professional issues.

The examples are presented in four sections:

- First year engineering
- Cross-disciplinary/professional practice courses
- Technical topic courses

- Capstone courses

The examples span a range of undergraduate student year levels, disciplines and national locations, as well as topics which are often claimed to be irretrievably gender neutral such as mechanics, thermodynamics and electronics. After the introduction, we have, where possible, structured each example to describe the relevant course features and then outlined the reasons for classifying them as inclusive. Each example concludes with what the authors tell us about the success of their innovations and strategies in improving their students' engagement with their studies. Where the gender of the students is not mentioned but we are told that *all* students improved, we have assumed that this means the female students as well as the males.

First Year Engineering

All engineering students need the same, or very similar, grounding in science, mathematics and information technology, and in many universities, students from all engineering disciplines follow a common, or nearly common, first year schedule of courses. However, there are some disadvantages in this arrangement. One is that students are taught in very large groups, sometimes making it difficult for many to engage effectively with their studies. Another issue is that a curriculum consisting solely of students' common curriculum content necessarily excludes material related to the field of engineering they have elected to join, and lack of information and enthusiasm about their future career is the reason given by many students for leaving during the first year. To address this problem, and at the same time encourage independent and interdisciplinary learning, several institutions now start the first year with a group project, in which students are required to explore the role and function of the professional engineer in various sectors of the economy, or suggest solutions to a contemporary engineering problem. These investigations are expected to go beyond the technical issues and incorporate ethical, aesthetic, social and environmental factors as well. Working in teams and developing the necessary communication skills adds not only to students' own skills but to their understanding of the work of a professional engineer.

Freshman Engineering Projects at the University of Hartford, US

Faculty at the University of Hartford were concerned about the scarcity of women in engineering and about the public image of an engineer, described as a '. . . boring individual, devoid of social skills who would rather spend more time with his electronic gadgets than with people' (Isaacs and Tempel 2001: 1). Faculty recognized that to increase the appeal of engineering to a wider range of students, including women, the opportunities for engineers to work in social and humanitarian fields needed to be highlighted. It was decided to introduce project work into the freshman year (the term for the first year at university in the US), which was designed to encourage students to explore how engineers contribute to the public good by, for example, providing safe water supplies and developing devices which help people with disabilities. Another essential element of these projects is the emphasis on the importance of good communication skills (Isaacs and Tempel 2001).

Topic Content and Teaching Approach

Classes are taught by two faculty members: one from Engineering and one from the Department of Rhetoric, Language and Culture, and each class includes students from electrical, computer, civil, acoustical and mechanical engineering. Twenty-five per cent of the students are women. For the group projects, students work in teams of four, deliberately mixed in terms of gender, ethnic origin and engineering discipline. Groups are changed for successive projects.

Faculty select project topics which will ' . . . maintain the enthusiasm for engineering that first semester freshmen start out with, and allow them to feel that they were 'doing engineering' before they had time to really learn to do anything' (Isaacs and Tempel 2001: 2). In the year reported (2001), three examples of the seven projects described are:

- Television Pilot. Teams of four students each devised a television episode where a gender and ethnically diverse team of engineers tackled a real engineering problem in a community setting. One object of the exercise was to depict the characters as having lives and interests beyond engineering. Students chose civil, environmental and biomedical issues for their plots and are reported to have enjoyed the opportunities it gave them to be creative.
- Design of a dachshund exercise track. The same class of students were re-grouped in order to have the experience of working with a different team. This time they were set the task of designing a dachshund exercise track, subject to given cost and land-use constraints, as well as the needs and concerns of various community groups. The teams had to present their design to the class orally, as well as submit a written report. The students were reportedly very amused by this project and enjoyed the scope available to work within the constraints.
- Cultural Studies: technology and society. The purpose of this project was to develop students' reading and writing skills and encourage them to consider multiple perspectives and appropriate use of different types of communications for different purposes and different audiences. Students studied various communications media used by, or referring to, engineers or engineering activity. Examples of written text ranged from Mary Shelley's *Frankenstein* (1831), to articles, corporate communications and poetry about technology. Students

also studied films about science (Isaacs and Tempel 2001), and we suggest that websites and social networking media could now be included. This project is assessed by three progressively more difficult essays addressing historical and contemporary engineering problems, including ethical issues.

Why is this Inclusive?

One of the stated aims of this model of freshman projects was to increase the appeal of engineering studies and the profession to women. By acquainting themselves with the body of knowledge about women's participation in engineering, or otherwise, faculty designing this course identified that many women engineering students want a career with a social and humanitarian focus and the opportunity to interact with people from all walks of life. These freshman projects are therefore designed to highlight these particular aspects of the work of professional engineers. By working in interdisciplinary groups, which are changed around, students get opportunities to develop team-working skills and interact with a wider variety of colleagues than those studying the same discipline. Finally, the cultural studies project helps students understand how engineering issues and perspectives can be influenced by cultural attitudes and, hence, to take account of the views and attitudes of others when devising technical solutions to industrial, economic and social problems.

How Successful Has This Been?

These projects have been successful in maintaining students' initial enthusiasm, resulting in 100 per cent retention in the year reported, when 25 per cent of the cohort was female. The groups of students from diverse engineering disciplines learned to integrate in order to work well as teams, thus developing collegiality and other skills needed in the professional arena. The cultural studies component helped to develop students' sensitivities to cultural issues and how these might impact on their learning environment as well as their future roles as professionals (Isaacs and Tempel 2001).

Current Update

Like many university courses, the one described in the 2001 paper we have cited has undergone some changes over the years. However, an examination of the information for the current version of this course, *ES 143 Engineering and Design*, taken by all freshman engineering students, indicates that the focus remains much the same and that projects continue to be related to disability or societal issues, for example the design of a stair climber. A group exercise is described about measuring the density of cake (then eating it!); it not only sounds fun but is also gender inclusive (Mellodge 2009).

Sustainable Engineering Practice at the University of South Australia, Australia

Sustainable Engineering Practice (SEP) represents a typical first year 'Introduction to engineering' type course that could be widely applied in the first year of the engineering program at many universities. Aspects of the design of this course have been discussed in Chapter 5 and we return to this example here to elaborate on the gender inclusive elements in the implementation of this course. SEP is taught to all first year students from across the range of disciplines of engineering offered at the University of South Australia (UniSA). These students represent a diverse group in terms of gender, race, language and cultural background and socio-economic circumstances. Although the primary delivery is face-to-face at a metropolitan campus, SEP is also taught at a rural campus where the students access lectures through online podcasts and meet weekly with their tutor via access grid (video conferencing) technology.

Topic Content and Teaching Approach

The aims and objectives of SEP, as well as the assessment and some aspects of the inclusive content, were described in Chapter 5. The course is taught via a one-hour lecture each week for the full class and two-hour tutorial sessions with a maximum of 25 students per class from mixed engineering discipline groups. SEP is co-taught by faculty members from civil and mechanical engineering (who are both female) with tutors from across a range of other engineering and non-engineering disciplines. The coordinators endeavour to ensure that a reasonable proportion of the tutors are also female (in 2009 this was 50 per cent). The project component, the *Engineers Without Borders Challenge,* has already been discussed in Chapter 5 as one of the inclusive aspects of the course content. However, the lecture component is also a core part of the inclusive approach in SEP, even though lectures might not usually be considered in this way. Lectures are primarily delivered by a range of guest speakers from industry who discuss a variety of sustainable engineering projects and topics, such as wind and solar energy, green building design and the principles of universal design for people with a disability. The range of speakers is distributed across all of the disciplines of engineering represented amongst the students in the course, which overcomes students' perceptions of lack of relevance to their own discipline area, which was one of the problems of common first year courses mentioned earlier. Other lectures focus on aspects of career development, a topic which helps to engage students by making their chosen career seem real to them, even though they are only in first year. In addition another lecture (delivered by one of the course coordinators) is specifically focussed on *Equity and diversity in the engineering profession.* As part of this lecture, a series of short videos of harassment and discrimination scenarios in university engineering classes is used to raise awareness and generate discussion amongst small groups of students within the lecture and to canvass possible strategies they could use to overcome such situations. Students are also advised of the support programs available to them at the university. In this way, equity is seen to be the responsibility of all staff and students at the university rather than just the people with a problem.

Why is This Inclusive?

The SEP course displays many of the characteristics of gender inclusive curriculum. A variety of teaching and learning methods and assessments is employed within the course, including small cooperative group projects with multi-disciplinary groups, opportunities for individual research and creativity through the student wiki and the encouragement of reflection on learning through the student blog. Lectures are available online to cater for part-time or remote students, as well as to allow review for those who attend classes. The teaching faculty and tutors are deliberately chosen to provide a number of female and non-dominant cultural role models. The course content itself is particularly gender inclusive since it involves projects and topics that consider social and humanitarian aspects of engineering and require students to develop culturally appropriate solutions. There is also considerable freedom for student groups to choose project topics in areas that interest them. Finally, the course also includes discussion that is specifically focussed on issues faced by women and other minority groups in the engineering profession.

How Successful Has it Been?

The SEP course has been taught twice thus far and, as noted in Chapter 5, is still undergoing modifications and adjustment in response to student feedback and faculty evaluations. Student reflections via their blogs and comments on course evaluations indicate that three of the course aims are certainly being achieved for the majority of students: the understanding of sustainability and sustainable engineering, awareness of engineering disciplines and the role of engineers and an appreciation of the social, political and economic contexts of engineering. Reaction to group work is mixed, and it is certainly a difficult issue for those managing the course since the groups are formed within the first two weeks of the first semester of first year, and hence there are inevitably some students who leave the program (not just the course) very quickly and groups must be adjusted. With regard to the performance of female students within the course, in 2009 there were 16 female students from 180 overall (8.8 per cent). The average mark for the female students was 10 per cent higher than the overall class average and their retention rate was 100 per cent compared with 90 per cent for male students, although both of these rates were pleasing for a first year course. Clearly such statistics will need to be monitored over a longer time period for a more complete evaluation of whether the course is successfully gender inclusive.

Cross-Disciplinary and Professional Practice Courses

The preceding first year course examples could also be considered under this category, as both involved students from a range of engineering disciplines and considered topics that are part of the professional practice criteria in accreditation requirements. However, the following two examples are particularly interesting because they extend these characteristics even further by either involving disciplines outside engineering, or students from a range of year levels, or both.

EPICS: Engineering Projects in Community Service

Purdue University in the US has extended the model of an introductory interdisciplinary engineering project in two significant ways. Firstly, the students work on current, real life projects. Projects needing a technical solution are proposed by community organizations, who become 'project partners'. The student teams work with their project partners to develop innovative technology-based solutions to real life problems, culminating in the acceptance of the proposed system and its implementation. Since each project must work through the whole process, from identifying the partner's needs, to the design, testing and implementation, a project cannot be completed in one semester. The need for project longevity is met by extending participation in these projects beyond the first year. First or second year students may join as new members of a team, and third and fourth year students may continue with a project or community partner that they have previously worked with, earning academic credit throughout.

Purdue created its *Engineering Projects in Community Service* (EPICS) program in 1995, initially for Electrical and Computer Engineering students, and has now extended it to students in all engineering disciplines, and also some non-engineering disciplines such as child development and nursing (Coyle et al. 2005). EPICS now operates in about 10 US universities and two in other countries. The primary purpose of EPICS is to give students the opportunity to work on multi-disciplinary, real life projects and to develop communication skills. The large scale and longevity of the projects and the vertical integration of the teams not only gives the projects the continuity needed for completion but also gives students opportunities to apply their increasing technical expertise and to undertake additional responsibility as they progress towards graduation. Other advantages of this program are developing effective communication skills with people from a range of social and educational backgrounds, learning teamwork across many disciplines and understanding '... the importance of professional and ethical responsibility, and the societal and global impacts of engineering solutions' (Coyle et al. 2005: 1).

Topic Content and Teaching Approach

The first stage is to identify appropriate projects and a not-for-profit community partner for each one, by inviting suggestions from faculty members, students and community organizations. The projects are then advertised to students, who are selected according to the discipline needs of the project. Teams consist of eight to 20 students, the large size enabling a range of disciplines and expertise to be represented, in order to achieve significant outcomes for the community. Teams meet for two hours every week to plan and progress the project and work on its technical requirements and challenges. All students in the program also attend a one-hour lecture by a guest expert on, for example, engineering design, communication or community service. Students may also opt to attend 'skill sessions' on topics such as computer programming or operating a lathe, if these are needed in their project.

Examples of projects undertaken in this program include the design and development of:

- computer-controlled toys for children with physical disabilities
- an environmental monitoring system to assist an art museum to improve preservation of its exhibits
- an outdoor high school science and environmental engineering laboratory
- hands-on science and engineering exhibits for an interactive science museum.

First and second year students earn 1 credit per semester, and third and fourth year students can opt to earn either 1 or 2 credits per semester. Students are assessed at both project and individual level on accomplishments, process, critical thinking, teamwork/leadership, and communication, in line with a detailed assessment schedule (Purdue University 2009b).

Why is This Inclusive?

The program designers are aware that certain curricular features are known to be successful in encouraging women to stay in science and engineering, and all of these are included in the program. These features are: '... framing science in its social context; stressing general educational goals, including communication, in engineering education; employing cooperative, interdisciplinary approaches; and undertaking problems with a holistic, global scope' (Coyle et al. 2005: 7). Another inclusive feature is that students are given opportunities to work with students from other engineering disciplines and from all years, thus providing them with opportunities to undertake a range of vertical as well as horizontal team roles.

How Successful Has it Been?

All students participating are asked for feedback through surveys that include open ended questions. *EPICS* students' comments from such surveys include:

- EPICS completely changed my opinion of engineering.
- It made me understand how every aspect of engineering (design, implementation, team work, documentation) come [sic] together.
- No longer is engineering just a bunch of equations, now I see it as a means to help mankind.
- Opened my heart (Coyle et al. 2005: 7).

Most students rank the experience with real world engineering design as their primary reason for participation, with many also ranking community service highly. Students also value the opportunities for teamwork and developing communication and leadership skills. A greater proportion of women enrol in EPICS than are represented in the various engineering degree programs. For example, between 1996 and 2001, women represented 10 to 12 per cent of enrolments in electrical and mechanical engineering programs but in the EPICS program women represented 20 per cent of the students in these fields. In addition, in the first three years of the program 20 per cent of EPICS students were women but 30 per cent of the team leaders were women. Coyle et al. suggest that with its emphasis on social issues, and its cooperative and interdisciplinary approaches, EPICS is succeeding in encouraging women into engineering and staying there (Coyle et al. 2005).

Current Update

The EPICS program continues to be active and is expanding to more and more universities around the world. More information is available from their website (Purdue University 2009a).

Rowan University Engineering Clinic

Rowan University in New Jersey, US, established its College of Engineering in 1996 and hence was able to develop a fresh and innovative approach to curriculum unfettered by the constraints of many years of former practice. Rowan University offers undergraduate degree programs in Chemical, Civil and Environmental, Electrical and Computer, and Mechanical

Engineering with a Bioengineering major also available. Key objectives of the program include:

- creating multidisciplinary experiences through collaborative laboratories and coursework
- incorporating state-of-the-art technologies throughout the curricula
- creating continuous opportunities for technical writing and communication
- emphasizing hands-on, open-ended problem solving, including undergraduate research (Rowan University 2009).

One innovative feature that has been adopted to meet these objectives is the *Engineering Clinic.*

Topic Content and Teaching Approach

The Engineering Clinic consists of a series of eight compulsory courses, one per semester throughout the engineering program. The courses are common to all of the engineering programs, and classes are a mix of students from the different engineering disciplines offered. In the first year the focus is on engineering measurements and reverse engineering, which is the systematic testing of existing products with the aim of trying to improve their design. Examples of products that the students have reverse engineered include coffee makers, hair dryers and electric toothbrushes. The second year clinics focus on engineering design in multi-disciplinary teams, with a significant emphasis on written and spoken communication in partnership with the College of Communication at the university. Past projects have included designing landfills and small bridges and undertaking energy audits of campus buildings. In the third and fourth years the projects are open-ended and are conducted in small teams under the supervision of one or more professors. Each project is unique and may run for multiple semesters, with most projects funded by industry or government departments. The projects require students to undertake a typical research/design sequence of literature search, problem statement, design and/or testing and final report in both written and oral form (Rowan University 2009).

Why is This Inclusive?

The Engineering Clinic's use of multi-disciplinary team projects based on real world objects commonly known to students, or real world design and research problems supported by industry, particularly with a social or environmental focus, clearly models an inclusive approach to engineering curriculum. The projects acknowledge prior experience of the students involved and address topics with which a diverse range of students can identify. The use of small project groups with faculty supervisors provides strong faculty-student mentoring, another feature that research has shown to support the retention and engagement of students, particularly women (as discussed in Chapter 3).

How Successful Has it Been?

Hartman and Hartman (2006) have analyzed the Rowan University Engineering program in some detail with regard to retention rates and reasons for staying in or leaving the program. They found that the program was more female-friendly than traditional programs. For the first five years of the program (1996–2001) the average retention rate for female, first year students was 85 per cent, slightly higher than the rate for male, first year students, which was 80 per cent. In comparing the percentage of science and engineering students who had left that field of study, the Rowan data from 1996–1998 was 31.4 per cent for males and 25 per cent for females, compared with national data indicating 41.2 per cent male and 35.7 per cent female. Across all retention measures (first year, second year and graduation rates) the Rowan data indicate that female rates were the same or higher than the equivalent male rates, indicating that the program is working for all students and particularly so for women.

The previously mentioned finding resonates with our repeated claim in this book, *viz* that a curriculum that is gender inclusive will also be inclusive for all students. Further evidence of this conjunction was found in the Rowan study as follows:

> The lack of difference between stayers and leavers in satisfaction with the applied aspects of the program—engineering clinic, teamwork and labwork—suggests that those features most female-friendly do not work to the disadvantage of the male students, i.e., they are as male friendly as they are female-friendly . . .
>
> (Hartman and Hartman 2006: 57)

This again endorses our claim that gender inclusive engineering education will work for the benefit of all students and of the profession as a whole.

Technical Topic Courses

Our research into gender inclusive engineering curriculum has consistently shown that some engineering topics are perceived by many faculty members as gender neutral, especially where the subject matter is solely technical and mathematical and hence perceived as impossible to adapt to be gender inclusive. Examples often cited are fluid mechanics, thermodynamics, electrical circuits and so on. In this section we have included examples of topics and courses which may be regarded as falling into this category. We have selected examples which are gender inclusive, at least in part if not overall. The gender inclusive characteristics arise either through the examples chosen, particular elements of the content, catering for a range of learning styles or through the teaching methodology, even though the overall course is still fundamentally technical in nature.

Electronic Circuit Theory

An inclusive approach to teaching electronic circuit theory has been developed at Jönköping and Linköping Universities in Sweden involving the extended use of 'problem-solving labs' (Carstensen and Bernhard 2009: 393) to improve students' understanding of transient response in RLC (Resistance/Inductance/Capacitance) electronic circuits. Their teaching approach is based in variation theory, a derivation of constructivist theory which emphasizes that experiencing difference rather than similarity improves learning. By working in laboratories equipped not only with the usual circuit boards and electronic components, but also with software tools, students are able to link the theories and models they are taught to objects and events with which they are familiar, in a more effective way than is usually achieved by traditional laboratory sessions taught separately from the mathematics required for a theoretical understanding.

RLC system outputs can be calculated either by solving second order linear differential equations, or much more easily by using Laplace transforms. However, successfully applying and manipulating Laplace transforms requires a good understanding of complex variables and complex functions, and traditionally, therefore, it has not been possible to teach this method until a fairly advanced stage in an engineering student's program. In addition, the only feasible way to examine the behaviour of RLC circuits under different constraints has been by varying the Laplace transform parameters; this requires considerable mathematical expertise, and the need to acquire these skills has for many students obscured an understanding of the behaviour of the physical system.

Topic Content and Teaching Approach

Working in the 'problem-solving labs' equipped as previously described, students are asked to investigate the effects on RLC systems caused by varying the parameters in Laplace transfer functions using tools ranging from pencil and paper to software such as MATLAB™, Simulink™ and circuit simulation. Firstly the students sketch the system behaviour manually for a simple case and then, using the software tools, they can generate accurate simulations comparing the variation in system behaviours under more complicated conditions. As well as the mathematical simulations, students construct physical circuits in which they can observe the effects of changing RLC values to verify the theoretical predictions. Thus, through examination of the behaviour of the system itself, students are guided to an understanding of the physical meaning of the parameters, without being diverted from this core task by having to grapple with advanced mathematical concepts and techniques.

Why is This Inclusive?

This teaching strategy recognizes that at the start of any class the students present will have a range of prior understandings of the topic and different learning preferences. The teaching described here is inclusive because it allows students to engage with RLC system output characteristics from a range of starting points and through different media. Some students, for instance, will find it helpful to consider the problem before handling any equipment, using sketches to anticipate the responses to various inputs. Some will enjoy using their IT (information technology) expertise in the simulations, and some will find it easier than others to conceptualize the problem mathematically. By working in groups in the laboratory, students can pool their strengths, knowledge and skills, helping each other to a better understanding of circuit behaviour from several perspectives. This is also an inclusive strategy because it enables students to connect mathematical techniques to a concrete situation immediately, as opposed to spending a long time learning the necessary abstract mathematics without understanding its relevance to the more practical parts of their studies. Students who have become disengaged from engineering studies have often blamed the difficulty of seeing the relevance of some of the curriculum to engineering. This latter point tends to be raised by women students who leave engineering, and thus a teaching strategy which keeps the relevance of the curriculum content well to the fore is viewed as gender inclusive.

How Successful Has it Been?

The student laboratory groups were videotaped and their conversations and actions subsequently transcribed and analyzed. Analysis revealed that, when the student groups encountered difficulties they turned, as intended, to other resources in order to resolve the difficulty and on the whole succeeded in establishing the necessary links. In cases where it was clear that students '. . . found making the intended links between the theory/model and object/event 'worlds' very difficult' (Carstensen and Bernhard 2009: 405), the teaching team made appropriate adjustments to the tasks set, the laboratory design and to the time spent in the activities. Thus, this course succeeded in relating the theory to the practical and in using active and cooperative learning methods, all of which are characteristics of inclusive teaching.

Engineering Materials

Engineering Materials is a compulsory course for all first year engineering students at the University of South Australia, which has been taught by the same faculty member for 10 years. When this faculty member was first assigned the course, he was challenged by his Department Head to respond to the previously poor student evaluations and results obtained in the course. His initial observations were that students struggled with unfamiliar terms and concepts and experienced difficulties in relating the theory to real life experiences. The faculty member became an enthusiastic participant in the inclusive curriculum project at the university at that time (see Chapter 7) and has continually developed and updated the course along inclusive principles since then, even though the content itself could be considered to be highly technical. In line with gender inclusive principles, the course employs a variety of teaching, learning and assessment practices, uses applications and examples from everyday life and maintains a focus on the environmental issues associated with engineering materials.

Topic Content and Teaching Approach

The course is structured around industry related case studies. Practical demonstrations are given in lectures of material failures, and displays of actual failed material components are used to introduce and reinforce the connection between theory and real life situations. Industry case studies are used as demonstrations and problem-solving exercises in tutorial classes to enable students to more readily relate the material to events in their own lives and to reflect on potential applications in their chosen discipline areas.

The environmental and human factor issues of engineering materials are also incorporated into the course. Examples include:

- the fracture of a marine oil pipe coupling at a refinery resulting in an extensive marine oil slick, to illustrate material behaviour and to allow reflection on the role of work practices and the engineering effect produced by divergence from standard operating procedures
- corrosion of steam pipes and pressure vessels with potential for serious, if not fatal, injuries thus emphasizing the association between materials performance, maintenance and management/financial decisions.

In a further example used as a tutorial exercise, students are asked to give a judgement on the responsibility for a fracture of a hired front end loader: is it that of the owner of a cracked component on the front end loader or that of the user who abused the component? The failure scenario is explained and fracture details discussed in class. In this example students are required to reflect on the evidence and suggest a judgement on the verdict and the distribution of damages using developed knowledge of engineering materials and properties. This approach encourages active involvement and engagement of all students in the class regardless of their prior knowledge. It involves ethical considerations, some knowledge and awareness of legal responsibilities and the capacity to set out a case clearly, both orally and in writing. Thus it is a good example of interdisciplinary engineering work that requires students to take on a professional role.

Engineering Materials uses a range of assessment methods and allows students to select different assessment strategies. Assessment includes online, in-class tutorial quizzes, written tutorial assignments and practical reports. If students fail to achieve 50 per cent on their continuing assessment (quizzes, practicals, tutorials), they will be required to sit the examination and the exam mark will replace the quiz mark in their assessment. Other students can choose to do the exam if they wish to improve their mark, but with a guarantee that their mark will not be reduced. This approach requires students to develop an awareness of their learning and to self monitor their workload and thus to take responsibility for their grade in deciding which form of assessment will best suit their particular situation.

Why is This Inclusive?

Introducing the course content by discussing the impact of errors and failures, rather than starting with abstract theory, enables the students to relate their studies immediately to the social and environmental aspects of engineering. This approach is likely to engage the interest of all (or at least most) of the students from the outset, whatever their prior knowledge. The flexibility in assessment gives students with diverse knowledge and skills the opportunity to demonstrate their engagement and learning.

How Successful Has it Been?

Student evaluations of the *Engineering Materials* course are consistently excellent, usually the highest of all first year engineering courses. When the faculty member first coordinated the course in 1999 he set a target of increasing the average mark for Engineering Materials to 55 per cent and raising the pass rate, by improving students' motivation through industry based examples and high levels of interaction in all classes. These targets have been realized convincingly: success rates have increased from 54 per cent in 1999 to an average in the high 80 per cent range over the last six years. Average marks have increased similarly from 48 per cent to 60 per cent, even though class sizes have more than doubled during this period to nearly 200 currently. Student evaluation comments consistently indicate a high level of satisfaction with the *Engineering Materials* course and show that the variety of techniques used in presenting the material maintains student attention and motivation. Gender disaggregated statistics for course success and retention are currently being investigated.

Engineering Thermodynamics

In Chapter 4 we described a perspective of the inclusive curriculum which views its purpose as encouraging students to challenge the power of the curriculum to support social injustice. This example describes how a second year *Engineering Thermodynamics* course is taught within a framework of 'liberative pedagogies' at Smith College, an all-women institution (Riley 2003). Riley describes liberative pedagogies as ' . . . an inclusive term, incorporating elements of critical/radical pedagogy of Paulo Freire (1971) and others, elements of feminist pedagogy, and the engaged pedagogy described by hooks (1994) that weaves these together' (Riley 2003:2). A key characteristic of liberative pedagogies is a student centred approach which is ' . . . directed toward creating a more equitable distribution of power in the classroom' (Riley 2003:2). Thus, students are encouraged to share with other students and their instructors the authority and experience they each bring to the classroom. Liberative pedagogies include many of the teaching methods we have already recommended in this and previous chapters, such as clearly stating learning objectives, making the content relevant to students' lives, facilitating active and cooperative learning and encouraging students to take responsibility for their own learning. Riley explains that, although some features of liberative pedagogies, such as problem-based learning, are shared with other pedagogies, what makes them 'liberative' is their potential to develop in students the principles and skills to make the engineering profession more accessible to all and to ' . . . empower future engineers to find or make a place in the field that resonates with their sense of social justice' (Riley 2003: 3). In other words this approach would engage with the political realities of engineering work and raise questions such as whose interests a proposed project might serve and who may be disadvantaged by it.

Topic Content and Teaching Approach

The bulk of Riley's second year *Engineering Thermodynamics* course covers the traditional basic concepts of thermodynamics ('equations of state and properties of pure substances; the first law; the second law; and the fundamental property relations' Riley 2003: 5) and their applications in a range of engineering disciplines. In addition, students explore issues such as the philosophical, historical and cultural contexts of thermodynamics. Non-western examples such as 'Chinese under-floor heating' and 'the development of water wheels and wind mills for power generation in the Muslim world' (Riley 2003: 10) are explored. There is an emphasis on relating thermodynamics to everyday life with topics such as energy balances in a pet turtle or ice-cream and 'phase equilibrium in playing the French horn, or in making the perfect snowball' (Riley 2003: 7). Students are encouraged to relate thermodynamics to engineering ethics, thereby engaging with aspects of the real world beyond the engineering issues posed by the particular problem.

Riley stresses the importance of matching assessment to course objectives: for example, in assignments students are encouraged to relate thermodynamics to their own experience and life stories. Riley points out the need for a range of assessment tools rather than an over-emphasis on tests which foster competitiveness at the expense of cooperation. For their group projects, students are required to assume a teaching role and speak authoritatively to the class about thermodynamics topics. As part of this project they must also develop materials, such as websites or a video, or give class demonstrations illustrating course topics. All of these elements form part of the broad course assessment.

Why is This Inclusive?

This example of a course developed by Donna Riley, a recognized protagonist and scholar of gender issues in engineering, spans the whole spectrum of understandings of gender inclusivity described in Chapter 4. It illustrates one way to ensure that all students are participating fully in the learning experiences offered. By making these learning experiences as varied as possible, a range of learning and technical skills is developed and rewarded. The course puts content in the context of students' daily experiences, considers social and environmental contexts and ethical concerns and promotes active and cooperative learning. It demonstrates the value of acknowledging and celebrating the skills and perspectives brought to the class by students from diverse backgrounds and, more unusually in the engineering literature than in some other disciplines, advocates helping students to develop critical awareness with a view to improving social justice in the profession.

How Successful Has it Been?

Riley gathers student feedback on various aspects of the course through surveys at the start, midway and end of the semester. As might be expected in a course which seeks to consider students' individual interests, strengths and weaknesses, different aspects of the course are ranked as favourites by different students. However, the group project, group test, class dynamics, relating of thermodynamics to everyday life and the historical components were five of the top six most liked aspects of the course.

Mechanics of Solids

As discussed in Chapter 4 the constructivist theory of learning states that all new learning is built onto prior experience and knowledge. Based on this idea, a model for effective learning called the '5Es', originally suggested by Atkin and Karplus in 1962, forms the framework for some imaginative teaching in mechanics suggested by Patterson (2008). The 5Es are:

Engage	capture the students' interest in the topic
Explore	give hands-on opportunities for students to explore and observe the phenomenon
Explain	provide theoretical explanation of what is happening
Elaborate	give students opportunities to apply what they have learned to new situations
Evaluate	test students' new understanding.

In the first of what will be a series of illustrated booklets for a project entitled *Enhancing Diversity in the Undergraduate Mechanical Engineering Population through Curriculum Change*, sponsored by the National Science Foundation in the US, the 5E framework is used to develop 12 comprehensive lesson plans and solutions for use by faculty teaching mechanics (Patterson 2008). The booklet, *The Mechanics of Solids*, is designed to assist students to understand the principles of fundamental mechanics and, by using examples from everyday life, to appeal to students from diverse backgrounds. The teaching strategy for all the sessions in this booklet is to structure each class on the 5Es. All of the topics presented in this booklet are introduced by the instructor in unconventional ways using everyday artefacts and materials. Two examples are:

- *Displacement plus deformation in control cables,* introduced by the instructor riding a bicycle into the class, then upturning the bike and demonstrating the gear mechanism by turning the pedals.
- *Conservation of energy/energy methods*, introduced by the instructor projecting rubber balls from a slingshot and asking students to identify the conservation of energy throughout this process.

The strategy for teaching another one of the 12 topics, *Stress and strain in uni-axial solid and hollow bars*, is presented in more detail below.

Topic Content and Teaching Approach

Engage: To grab the student's attention, the instructor takes an iPod into the class and dangles it by the earphone cable. An old cable is then cut to expose the copper wire and insulation.

Explore: Lengths of copper wire and strips of empty insulation are passed around the class for students to examine and stretch to encourage discussions about relative extensions and stress.

Explain: Instructor works through numerical examples in which the mass of iPod, the diameter and length of wire and the external diameter of the plastic insulation are given. Assuming in turn that the wire and the insulation carry the entire load, calculations of both the stress and extensions in each case are calculated, finding that these differ for the two materials.

Elaborate: Students are now asked to consider the real situation: which is that in practice the wire and the insulation will bear the load together and will necessarily extend by the same amount. The class discusses what actually happens to the tension in each, and these are then calculated.

Evaluate: Students provide evidence of their learning by calculations of extensions in the cables of a suspension bridge when a truck crosses it, stress on a femur of mammals of various sizes and tension in cello strings (Patterson 2008).

Why is This Inclusive?

In line with the constructivist approach, the phenomena to be explored (stress and strain in different materials) are presented in the context of an iPod, a device which can safely be assumed to be currently well known, popular and of interest to the vast majority of any group of students. By handling separately each of the materials which form the cable, students are invited to anticipate what will happen if each one bears the weight of the cable separately: that is, to apply their own experience to understanding the problem. The instructor gives the students time to consider the problem before providing the explanation and then the methods of calculating the exact solutions, thus making allowance for the variation in students' prior knowledge. At the *Explore* stage, students are encouraged to discuss with each other how the tensions in each material are affected when, as happens in reality, the two materials are constrained to stretch by the same amount. This process ensures that all students are actively involved in the learning experience.

How Successful Has it Been?

The example provided here is one lesson within a series of 12 that have only just been released. Hence evaluation of their effectiveness has not yet taken place. Further booklets are under development for other fundamental mechanical engineering course areas such as thermodynamics. Further details are available at the project website (EngineeringExamples.org 2009).

Physical Electronics

In the second year *Physical Electronics* class at the Malaysian campus of the Australian based Curtin University of Technology, the nine female students in a class of 30 always clustered together and sat apart from the male students. In marked contrast with the male students, the women appeared to be extremely reluctant to participate in the class by answering questions. When they did respond, this was more likely to be in reply to questions that tested memory rather than to questions requiring some analysis. These factors concerned the faculty member teaching the course, who interpreted the way the female students all sat together as signalling discomfort (Stäuble 2004). Additionally, she was aware that, by not participating, the female students were losing some learning opportunities. Such silence has been described as 'a position in which women experience themselves as mindless and voiceless and subject to the whims of external authority' (Merriam and Caffarella 1999: 146).

Initially Stäuble tried to remedy the situation by addressing some questions to the female students directly, but she abandoned this strategy when the female students still refused to answer and she realized the danger that the male students might feel discriminated against and the female students might feel that they were in the spotlight (see Chapter 3). Stäuble therefore sought other ways to encourage the female students to participate by offering additional auditory, visual and kinaesthetic learning opportunities. Two of the three activities she successfully trialled are described next.

Topic Content and Teaching Approach

The first example is to demonstrate diffusion of electrons within a solid, and the second one is to explain the concept of a 'hole', in which the absence of an electron is treated as equivalent to the presence of a hole.

To demonstrate the process of diffusion, perfume was sprayed at the front of the classroom and students were asked to raise their hands when they were able to smell it. The class then discussed the pattern of diffusion observed and were invited to predict from this the pattern of electron diffusion in solids.

A more kinaesthetic activity was used to explain the concept of an electron hole. A row of eight chairs was set up with the left end designated negative, and the right end positive. Seven students were asked to sit on the chairs, leaving the right-hand chair empty. After an explanation of the Pauli Exclusion Principle and its impact on electron conduction, students were asked to model the movements of the electrons in relation to the hole, by moving to occupy the chair on their left when it became vacated. They thus observed that the empty chair successively 'moved' towards the positive node (Stäuble 2004). This observed behaviour was used to explain that the rightwards movement of many electrons can be regarded as equivalent to the leftwards movement of a single hole. This is a useful simplification, because the movement of a single entity (even if it is fictional) is much easier to analyze when modelling diffusion than is the movement of many entities (even though they are real). In both of these examples students were required to exercise physical presence in the classroom (at least in terms of those chosen for seating in the second example), and thus the activities became embodied issues rather than a series of computations or paper based diagrams. They had to be breathing, sensing, moving, observing and recording. Thus their learning was an active involvement, rather than the passive one of just being told.

Why is This Inclusive?

This example is gender inclusive because in the first place the faculty member noticed the non-participation of the female students and recognized that they were in danger of being excluded from the full range of learning opportunities offered. She then explored ways to encourage the female students to engage with the class activities to ensure that they were included in the learning experiences provided. To do this effectively, she sought guidance from the literature in this area and experimented until she found a successful solution. One of these solutions was to illustrate a physical principle using perfume, a substance likely to be familiar to the women. Another was to devise active learning activities which required all students to mingle in an active way and then discuss the analogy between their activity and the movement of electrons.

How Successful Has it Been?

Presenting the technical material in a different way, and requiring the students to physically move around, was successful in overcoming the problems of the self segregation of the female students and their reluctance to participate in the class by answering questions. In all the activities described, it was observed that the female students participated to the same extent as the males and also appeared more comfortable in this more student centred environment than they had previously. The female students' observed delight with the first activity was clear evidence of their engagement. (But if any readers would be reluctant to use perfume, we suggest that the same effect can be achieved using any of a number of household spray products.) Additionally, all students were asked to complete evaluation questionnaires, answering questions about their views of these new learning activities. The results showed that the female students ranked the fun and interest of these activities more highly than did the male students and, likewise, the women claimed that the activities assisted their understanding more than the men did. However, feedback on the value of the group interaction, and on all other questions, was roughly equal between the women and the men. The conclusion was that the activities succeeded in increasing female participation but did not disadvantage the male students at all.

Capstone Courses

Most engineering programs include a 'capstone' experience in the final year, which is intended to bring together all of the knowledge and skills gained in the earlier years of the program into a realistic and integrated design or research experience. Some countries' accreditation conditions make this a compulsory program component. Capstone courses can take a range of different forms and account for a varying amount of credit in a program. In some cases they are undertaken by individual students and in many cases by small student teams. Some are interdisciplinary and others are not. Some capstone courses focus on design experiences, others on research and some on a combination of both. Two Australian examples are provided here, each of which describes a capstone course that differs from many others. In the first case the capstone course is a whole class design project, and the second example describes a course where engineering students work with business or community partners together with students from non-engineering disciplines.

Civil Engineering Design Project

The *Civil Engineering Design Project* is a compulsory capstone course taken in the final year of the civil engineering degree at the University of South Australia. It forms 25 per cent of the final year student load and has been designed from the outset to be inclusive. The 14-week course is focussed around a group learning experience, as the whole class works as a self-managed design consultancy for the duration of the project.

Topic Content and Teaching Approach

Each year, a real civil engineering project in the early stages of planning or construction is identified, and a partnership established with the relevant industry group managing it. Through this partnership, a suitable project brief is developed by the industry partner in conjunction with the faculty members coordinating the course, and the required permissions are obtained for the students to have access to any relevant site information such as geotechnical reports or environmental assessments. Examples of recent project briefs include the redevelopment of Adelaide Airport and a housing development located on a site that was contaminated by industrial usage.

Mirroring the processes seen in real construction projects, students are then asked to develop and present tenders (stage 1), feasibility studies and concept designs (stage 2) and finally detailed designs (stage 3) for the project brief given. Except for stage 1, which runs for the first two weeks, the class functions as a single company. Company and team group leaders elected by the students form the management team. This team sets work programs, allocates personnel according to the specific needs of the project and the project stage, is responsible for communication between teams and collates final documents. The composition of any work team is carefully monitored by the faculty coordinators in consultation with the student management team to ensure that the class develops structures and groups which support students from social or cultural minority groups or those with English as a second language. Leadership and task roles are rotated in each stage of the project to ensure that the majority of students can take on positions of responsibility at some stage of the project. Each student must keep timesheets which are audited by the Quality Manager. Students are also required to keep a reflective diary. All students are required to make an individual oral presentation to the class at the end of one of the project stages. The Project culminates in a public presentation of the work to students' families and invited members of the engineering profession. Speakers at the public presentation are volunteers from the class.

Guest lecturers from industry are used to present on topics such as quality management and contract processes. An informal lecture/class discussion program led by the faculty coordinator also addresses topics such as working overseas, team skills, recognizing different personalities and contributions of individuals and the issues facing women in the engineering profession.

Assessment consists of five components. A class mark, a group mark and a peer assessment attract equal weightings of approximately 30 per cent each, with the remaining 10 per cent divided between the oral presentation and reflective diary marks. Students assess members of their own groups in terms of their contribution to the team, usually displaying a responsible and mature attitude to these peer assessments. Marks are finalized only after informal discussions have taken place with all class and group leaders, as well as individual students identified as possibly disadvantaged, or as requested at any time by any student.

Final year cohort numbers and hence company sizes over the 12 years that the course has been run by the current coordinator have ranged from 30 to 60. Whilst the project scope can be adjusted to suit class size, the optimum number for a single company is about 40, so in the future, larger cohorts may be split into two companies.

Why is This Inclusive?

The project requires student teams to propose a solution to a topical and realistic problem in an industrial context which also has environmental, social and economic implications. Management of the teaching and learning environment to promote inclusiveness is recognized as critical. The course structure promotes teamwork and cooperation and encourages a large number of students to take on leadership roles. The assessment approach allows for significant student input, and students also have considerable flexibility in terms of nominating which roles and areas of work they would like to undertake (this is done on a first, second and third preference scheme, and a student who does not get his or her first preference in one stage is guaranteed it in the next stage). Issues relating to working in teams, professional ethics and the position of women in the engineering workforce are also explicitly discussed in the class seminar series.

How Successful Has it Been?

The *Civil Engineering Design Project* is evaluated annually through the use of the standard university course and lecturer evaluation instruments and at various stages has been more intensively evaluated through focus groups and student surveys. The percentage of students who have agreed or strongly agreed with the statement 'Overall I was satisfied with the quality of this course' has ranged from 91 per cent to 100 per cent over the last seven years. Text responses to questions regarding what students have found most useful for their own learning overwhelmingly focus on the 'real life' aspects of the project and the professional skills development such as teamwork, communication, leadership development and so on. For example, 'This course offered not only real-life project experience, but was structured so that the emphasis was on individuals within the team working together.' Gender disaggregated data are not available from the anonymous evaluation instruments and the student retention rate has been 100 per cent for both genders for many years, which is not surprising for a capstone course just prior to graduation. However, women are regularly represented in higher percentages in leadership roles in the project than their overall percentage in the course, which indicates both their confidence and comfort in undertaking such a role within the learning environment of this course and the support of their class peers in electing them.

Greenhouse Gas Emissions

Final year Bachelor of Engineering (Mechanical) students at the RMIT University, Melbourne, Australia, can opt into the *Greenhouse Challenge Support Program*, a collaborative project between RMIT and local public and private sector organizations which receives some financial support from the federal government. The project has been run since 2004 as an option that can be chosen to satisfy the capstone course requirement for mechanical engineering students, who undertake a 12-unit project course in each semester of their final year (making up 25 per cent of their course load that year). The projects explore how the host organization can reduce greenhouse gas emissions. In 2007 additional funding was obtained, and the project was extended to include multi-disciplinary student teams.

Topic Content and Teaching Approach

In the first three years of the project's operation, individual mechanical engineering students undertook the projects under the supervision of a faculty member in conjunction with an industry organization. From 2007, students have worked in multi-disciplinary teams of three consisting of one Mechanical Engineering student and two team members drawn from Social Science, Architecture and Design, or Management programs (Bunting et al. 2007). The course aims include providing students with the opportunity to work on a real life project on an issue which is of wide societal concern and to broaden the type of investigation undertaken by students beyond the purely technical perspectives. Student teams work with a company to make recommendations on increasing energy efficiency and reducing waste, taking into account the social and environmental implications of their recommendations as well as the technical ones. Project topics are developed by the faculty members in liaison with the participating companies. Examples of the topics in 2007 were:

- investigating the feasibility of small-scale wind generation in a new urban development
- reducing energy usage in student computer laboratories through redesign of lighting and use of computer power saving on a university campus
- designing technical and behavioural initiatives to reduce emissions that were in keeping with customer expectations of a five star hotel.

(Bunting et al. 2007: 2)

The teams work under the joint supervision of an industry supervisor from the hosting firm and a faculty member from RMIT. Each student also has a discipline-specific supervisor. The teams are expected to meet weekly, working at the hosting company when appropriate. Students receive preliminary training in topics such as teamwork, project planning and greenhouse gas accounting. Because the engineering students' program requires their project to be two semester-long courses, while the other disciplines' projects are single semester courses, the projects have to be structured into two halves. The whole team therefore works together for the first semester, and the engineering student completes the project in the second semester.

Why is This Inclusive?

Like several examples given earlier, this course is inclusive because it provides opportunities for engineering students to gain experience in working with colleagues from other fields, as they will almost certainly need to in professional life. The focus on a critical issue facing both industry and society in general in a real context and from a variety of perspectives adds to the relevance of their studies to their chosen career, fulfilling a need frequently expressed by students, often female, who become disappointed with engineering studies.

How Successful Has it Been?

The authors have formally evaluated the multi-disciplinary teamwork aspects of the course, using semi-structured interviews, and found that students' skills in project management, interdisciplinary communication and conflict resolution were strengthened. In addition, success in exposing a wider range of students to the opportunities to work on sustainable energy projects was demonstrated not only by the involvement of students who had expressed an interest in this area, but also by the number of students who had not initially shown any interest in this field but whose interest was sparked by these projects. This course therefore appears to have succeeded in ensuring the active participation, and broadening of experience, of all students, including the women (Bunting et al. 2007).

INCLUSIVE CURRICULUM SNIPPETS

During our search for course exemplars we also came across many excellent gender inclusive ideas that could be applied to some parts of the content of a course; or to the teaching and learning arrangements used. Sometimes we found description of courses that appeared to be gender inclusive but had insufficient information available for us to include them as full course exemplars. We have therefore structured them in similar categories to the course exemplars and hope they may be useful in sparking ideas about something you could apply to your course or program to make it more gender inclusive. Many of these snippets are presented as quotations from publications that also may provide other such examples for you to peruse.

First Year Engineering

Boise State University has worked on several strategies to integrate math with engineering applications in the freshman year. . . . Eight modules have been developed; the one people most frequently request information about is "Peanut Butter Cracker Manufacturing: Overall Design, Testing, and Implementation of All Facets of a Manufacturing Assembly Process (Plumb and Reis 2007: 27).

A freshman/sophomore course at the University of California at Berkeley called *Designing Technology for Girls and Women* encouraged students to examine gendered perceptions of the design process (Agogino et al. 2004). 'The course covered gender issues associated with new product development from a human-centered design perspective. Students worked in multidisciplinary design teams and participated in interactive workshops with target users and industry sponsors. . . . Evaluation showed that students developed a strong belief that 'good design' dictates that technology can and should serve all members of the potential user population, including those traditionally underrepresented with technology' (Dym et al. 2005: 107–108).

Cross-Disciplinary/Professional Practice and Capstone Courses

The Colorado School of Mines has developed a minor in humanitarian engineering. So far, the proportion of women participating in humanitarian-engineering design projects exceeds the proportion of women in the general population of engineering students. Completed designs include an orphanage in Romania, a drip-irrigation system in Senegal, a water-and-sewage system in Honduras, and village lighting in Ecuador. As a result of the program, students sometimes change their career plans. For example, instead of working for a microchip company, one participant wants to use her electrical engineering degree to work on power distribution for people in developing countries (Plumb and Reis 2007: 27).

The University of Notre Dame also offers undergraduate research experiences in humanitarian and aid projects in developing countries. Nine out of 10 participants are women 'despite physical hardships and lifting requirements' with a similar ratio in subsequent years (Busch-Vishniac and Jarosz 2007: 256). This book chapter by Busch-Vishniac and Jarosz provides many other examples of inclusive curriculum snippets and is well worth reading.

A 'diversity lecture' was incorporated into a fourth year course *Management and Human Factors in Engineering* that was compulsory for all engineering students at the University of Wollongong in Australia. It was presented by a range of guest speakers including a female pro-vice chancellor, the director of equal opportunity at the university, a female engineering lecturer and postgraduate student and three industry based engineers (two female and one male who was married to a female engineer and who talked about juggling two engineering careers with children). Readings and an assignment on the topic was also set. Feedback questionnaire results indicated a significant increase in awareness of diversity and gender issues for most students (Schafer 2006).

Technical Topic Courses

Dr Neil Abraham of Bryn Mawr College designed a program to keep undergraduate women in science. 'In addition to designing introductory courses with a minimum of prerequisites so that young women who had not taken physics in high school could enrol, he also involved his students in talking and writing about physics, drawing on their verbal as well as quantitative skills (Kahle 2002: 14). He thus took advantage of their verbal skills rather than limiting his approach to traditionally masculine hands-on tinkering and competition, or to memorizing' (Busch-Vishniac and Jarosz 2007: 251).

Workshop Physics is a scheme developed at Dickinson College and Tufts University in the US. 'An unusual feature of WP [Workshop Physics] is its emphasis on kinesthenic experiments. For example, students run after a ball rolling along by itself. Then they repeat the experiment, hitting the ball repeatedly with a stick, finding that the second ball is much harder to keep up with as it goes faster and faster. This experiment clarifies the difference between motion at a constant speed and motion at constant acceleration.' (Whitten and Burciaga 2000: 219).

INCLUSIVE CURRICULUM SUGGESTIONS CATEGORIZED BY ATTRIBUTES OF ENGINEERING GRADUATES

In Chapter 5 we used the *Attributes of Engineering Graduates* that we developed (Appendix A) as a potential framework for inclusive curriculum design. This can be particularly useful in universities that are required to use a similar framework for program design or where programs and courses must be related to accreditation framework that use similar attributes. In this section we provide a range of additional suggestions categorized in the same way but focussed more at the course level.

Body of Knowledge

- Make the history of engineering part of the program; for example, in an early year course provide an historical overview of the development of engineering from 6000 BC, focussing on the problems which needed a solution.
- Acknowledge the different approaches to the construction of scientific knowledge: from the 'traditional' objective approach typically adopted in western curricula, to the 'connected procedural knowledge' of some other cultures which seeks understanding and meaning from the individual's perspective.
- Ensure that women's contributions to technological development are also discussed (for example, Hedy Lamarr, a film star, who patented a coding technique now widely used in telecommunications).
- Use examples of general familiarity and interest (for example, health related devices for the disabled; sanitation), electricity and electronics in everyday contexts (for example, domestic or musical).
- Provide introductory 'how-to-use' laboratory and computer sessions as an integral part of the program, open to all students, to ensure that those who have had limited access to computing facilities, or who have never had the sort of experiences often assumed, like playing with mechanical or electronic toys, acquire necessary or desirable prior experience.

Lifelong Learning

- Set open-ended laboratory investigations or computer simulations, as well as exercises with 'closed' expected results.
- Ask students to reflect on experiences in the classroom or in undertaking assignments, in writing. This provides a 'safe' medium for all students to participate without being too publicly exposed and also helps develop language skills.

- Be aware of the importance of 'attribute style' (the pattern of causes attributed by people to the events of their lives) and the known gender differences in these, such as women students attributing failure to lack of ability whereas male students may 'blame' bad luck (or even bad teaching) (Beyer 1991).

Problem Solving

- Consider the place in the curriculum of systems and problems which involve uncertainty.
- Set engineering problems which address a societal need, rather than focussing only on technological solutions. For example, in the course *Mechanics and Structures* at the University of South Australia, students were asked to design a structure for getting disabled patients in and out of swimming pools for therapy.
- Consider the importance in the curriculum of students' development of the skills of sound judgement when faced with practical or ideological choices.
- Consider the relevance and the appropriateness of the context of assignments and exam questions. Students unfamiliar with 'western' male sports, for example, are at a disadvantage if asked to solve problems set in such contexts.
- Make creativity and innovation an integral part of the program. For example, the course *Engineering Design and Innovation* at the University of South Australia combines a problem-based approach to engineering education with techniques from the field of educational psychology to develop creativity and innovation in students.

Teamwork

- Use a variety of strategies to organize groups within the same course or the same program:
    - in some cases faculty members should deliberately select the groups and try to distribute the (few) females, the non-English-speaking background students and international students evenly around groups to ensure differences in approaches, backgrounds, opinions in the groups. Also consider mixing personalities, sometimes deliberately putting dominant personalities together to encourage them to learn to work together.
    - in other cases allow students to select their own groups or adopt a compromise between these two strategies.
- Use peer assessments of individuals within groups and of groups by other groups. Ensure that these assessments are seen as important by making them a significant component of assessment.

- Provide short talks and exercises on teamwork, personality differences, negotiation skills to assist students' understanding of group dynamics, both in introductory courses and revisited in later courses.
- In *Mechanics and Structures* at the University of South Australia peer assisted learning sessions or study groups are used (held separately from, and with a different purpose to, faculty-led tutorials):
  - o led by successful students from previous years who are interested in participating in this way
  - o current students indicate that they value the sessions for explaining topics in language they understand, reinforcing concepts, etc.
  - o student leaders value the mentoring and personal growth experiences the role provides for them.

Ethics

- Discuss the impact of technology in indigenous cultures so far, and its potential for the future.
- Discuss the place of human values in engineering in relation to the situations of uncertainty and instability arising in practice.
- Examine the role of engineers as agents of change and social experimenters (do they always take into account the values and interests of different environmental and community groups?).
- Focus not only on the technical solutions of engineering problems, but also on the purpose and the social benefits.
- Discuss the ways in which technology has improved people's lives.
- Include real life examples where students are required to consider the ethics of issues and incidents arising in engineering.

Communication

- Extend students' skills beyond the modes of consultation of the scientific discourse to those appropriate for the professional and social contexts of organizations and markets, the community and the environment.
- Have students role-play situations: for example, community consultation meetings between the transport department and residents whose houses are going to be acquired for a new freeway.
- Provide multiple opportunities throughout the program for students to present their work in multiple modes: orally, written reports, posters, conference papers.
- Provide students with the opportunity to present their work to members of the industry, their families and the general public: for example, final year project presentations.

- Consider the cultural formation of the engineering profession and the implications for its future.
- Be aware of, and accommodate, culturally influenced classroom behaviour: for example, allow students plenty of time to respond in class. Not everyone wants to respond spontaneously; some prefer to compose their response before replying.
- Challenge sexist or racist language or behaviour arising in the classroom, laboratory or workshop.

Global Context

- Discuss the role of technology in third world cultures.
- Arrange overseas student placements if possible.
- Encourage and facilitate 'virtual' collaboration on projects with students in overseas universities.

FREQUENTLY ASKED QUESTIONS

We close this chapter with some of the questions that we are often asked by faculty members who are concerned about issues of gender inclusive engineering education. As with many questions, there are not always simple answers, nor answers that will be the best for all situations at all times. But we hope that these FAQs will be of some assistance when you are next wrestling with one of these questions in your course.

Do I Put Female Students in the Same Team or Split Them Up?

Example question:

There are four female students in my class of 40. When working in groups, should I put them all in the same group so that they are not dominated by the male students or split them up to make the groups as diverse as possible?

Answer:

Although this seems a simple and straightforward question, unfortunately there is no simple and straightforward answer. There are advantages from the gender inclusive viewpoints in both types of groupings, and if we were talking about inanimate objects the best strategy might well be to alternate between the two types of groups. But it is important not to make assumptions about individuals' preferences for team composition, though we should always be aware that these may differ from our own. For example, we should not assume that the four girls get on together and would work effectively in the same group. They also might resent being allocated in any way according to gender, on the grounds that they are not aware of any existing gender issues in the class and do not want to trigger any. For laboratory work, Rosser (2004) advises that females should always be paired with females, to avoid the often observed situation that in male-female pairs, the male does the experiment and the female records the outcomes. Returning to group work, our advice would be to initially rotate group memberships in such a way that apparently 'accidentally' sometimes the women are all in different groups, sometimes there is a group with two women, and sometimes three. However, before arranging things to get all four women in the same group, we would advise it is probably worth discussing the matter with the class, using the opportunity to give them some background of the known gender issues in engineering education and employment. Tell them about Rosser's research which found that all-women and coeducational groups performed best in a physics class. The all-men groups performed worst because they were 'jockeying for position' (Rosser 2004: 37). So from that point of view of best performance, perhaps it's a good idea to have all women or at least two women together. Hartman and Hartman also found from analysis of the Rowan Engineering clinics that 'Women who shared the teamwork experience with at least one other female over the course of the first year were more likely to become more positive in their attitudes toward group work' (Hartman and Hartman 2007: 89). On the other hand, this situation would almost never happen in the workforce, and perhaps from that point of view it is best to have one woman in a group of men, since that is likely to happen to most women when they are in the workforce. Hence the answer to this question must be worked out in the particular context that applies to those women in that course.

There is One Woman in My Class. How Can I be Inclusive?

Answer:

Your sole woman student will almost certainly feel uncomfortable at times, though this may not be apparent. (An analogy is to imagine being the sole father taking your child to a playgroup gathering offered at your local community centre, where you are not included in the informal conversations of the mothers in attendance about child rearing, even though it is a discussion to which you could contribute, and you are also excluded from discussions about topics that are unfamiliar or of no interest to you. At the same time, you are viewed with suspicion by some of the mothers about your motivations for being there, or you are seen as a bit odd, even though you are as passionately interested in bringing up your child as they are and you have every right to attend the playgroup.) So, firstly, keep a kindly eye on the sole woman without drawing attention to her in any way and prevent the atmosphere from becoming too 'blokey' (from the Australian vernacular, somewhat difficult to translate, but think of the type of atmosphere at a party with lots of males talking about football and cars). Do not accept any overtly sexist or discriminatory comments or behaviour from any member of the class, but ensure that you do not put the sole woman in an awkward situation in the way that you deal with this. Do not single the sole woman out for attention either, for example with an opening that all female engineers hate:

'Gentlemen . . . and lady'. Finally, remember that the evidence shows that gender inclusive teaching methods benefit all students and so it is still appropriate to use the gender inclusive curriculum examples recommended here.

I've Tried to be Inclusive But the Female Students Are Not Attending

Example question:

I have tried very hard to be gender inclusive in my teaching. For example, I have catered for women's interests by replacing my demonstration of the workings of an electric drill with a vacuum cleaner—a machine that all women are going to make considerable use of throughout their lives. Yet the female students are still not attending my lectures. Why not?

Answer:

All the evidence says that engineering students want to be interested and excited by their future careers, to motivate them to stay with their study. And yes, faculty are advised to relate the curriculum content to students' everyday experiences to achieve this engagement. But it is risky to make assumptions about what devices and machines women would identify as interesting. Ask women that you know whether they regard a vacuum cleaner as an interesting artefact, or a washing machine; or whether the typical male student is interested in the workings of a lawn mower, for instance. It is important not to move from making one set of assumptions that exclude gender considerations to another set of assumptions that incorporate stereotypes of women. Please peruse the many examples provided in Chapters 5 and 6 to consider alternative approaches.

But the Stuff I Teach is Very Technical; How Can I Make That Gender Inclusive?

Answer:

This is probably the most common response that we receive to any mention of gender inclusive curriculum approaches. We do not dispute that a differential equation is not gendered, that the first law of thermodynamics does not change, that every action has an equal and opposite reaction and so on. However, we hope that if you read through the discussion we have presented in Chapters 1 and 4, as well as the guidance and many examples that we have provided in Chapters 5 and 6, you will realize that curriculum comprises far more than the content. Technical content can still be taught in ways that are gender inclusive, by considering the learning environment, the forms of assessment, the methods of teaching and more. And even if the fundamental theory doesn't change, the examples that are used to illustrate it, the projects that can be utilized to assist students to make their own discoveries about it to further develop their understanding, can certainly be made more gender inclusive than those traditionally used in the engineering classrooms of the past.

SUMMARY

In this chapter we have provided numerous concrete examples of gender inclusive curriculum, from first year to capstone courses as well as examples of courses that are often considered incurably technical and impossible to change. We have included some 'snippets'—cues to spark more ideas as you think about the particular situation of the courses you teach in the context of your university and the diversity and gender mix of your classes. We have answered some of the most common questions we are asked about gender issues in engineering classrooms. Chapters 5 and 6 have focussed on what individual faculty members and faculty teams can do to make their courses and programs more gender inclusive. In Chapter 7 we look at what is needed to embed gender inclusivity into university and accrediting body policies and practices to ensure its long-term sustainability.

7 Implementing and Embedding Gender Inclusive Engineering Education

INTRODUCTION

The previous chapters have considered the processes and issues involved in introducing inclusive features in the engineering curriculum at classroom or course level, and designing an inclusive course or program at departmental level. But without declared and monitored institutional commitment to inclusivity, there is no guarantee that the next instructor to teach the course, once the inclusivity champion has moved on or retired, will continue with inclusive practices in the classroom. Likewise, unless institutional procedures evaluate course processes against their declared objectives, there is a real danger that objectives are ignored if the course manager and instructors have no particular commitment to inclusivity. The same argument also applies to the non-curriculum based initiatives that were discussed in Chapter 3, which we also view as a critically important component in the overall approach needed to implement gender inclusive engineering education.

In this chapter, we consider the policies, strategies, processes and systems needed to ensure that gender inclusivity is embedded into an institution, with particular emphasis on engineering. These recommendations are illustrated with three case studies from universities or university coalitions which have achieved gender inclusivity in some measure in at least some areas of practice (but there is always room for improvement!). We then consider the need to embed gender inclusive practice into professional accreditation processes and policies: a key means of influencing the practice of almost all engineering departments, regardless of geographical location and institutional characteristics. Finally we discuss the need for evaluation of any programs or measures that are implemented and offer an example of one measure that can be used to do this.

EMBEDDING CHANGE IN THE STRUCTURE AND CULTURE OF ENGINEERING DEPARTMENTS AND INSTITUTIONS

An inclusive curriculum can only flourish in an inclusive university. Seymour, reflecting on the findings from her major 1997 study with Hewitt on why women leave engineering (Seymour and Hewitt 1997), stated that the main reason that efforts to retain women and minority students in engineering were not more effective was that 'they supported individual students in an unremediated educational context' and paraphrased Einstein to note that 'you cannot resolve a problem in the conditions that created it' (Seymour 2002: 81). Fox et al. in their evaluation of most and least successful women in engineering programs emphasized that a key component of the most successful programs was the attempt to '*adapt* the institutional environment to broader programmatic aspirations' (Fox et al. 2009: 347) by being more actively engaged with influencing institutional directions and developing more extensive links with faculty and administrators. Less successful programs tended to '*adopt* the goals of the institution' (Fox et al. 2009: 347) without challenging the organizational structures of educational programs. The US National Academy of Sciences (NAS) also noted that one of the common threads in their study on recruiting and advancing women students and faculty in US science and engineering was that the success of such efforts depended largely on whether 'university leaders and administrators promote the institutionalization of change, not quick fixes, and on rapid implementation' (NAS 2006: 8).

These and other studies all point to the need for institutional commitment to occur before sustained success will be achieved in improving the access and retention of women in engineering. Too many strategies and programs to date have relied on the efforts of individuals or small groups. Such contributions are essential, but without institutional commitment and embedding in policy and practice they are unsustainable over the longer term when personnel, priorities and budgets change. A range of strategies or areas must be implemented or addressed across institutions if sustainable change towards gender inclusive engineering education is to occur. These include:

- Visible institutional leadership
- Equity/diversity policies and practices
- Inclusive culture and curriculum developments and embedding
- Faculty development
- Inter-department and cross-institution collaboration
- Coordinated women-in-engineering programs or similar initiatives
- Ongoing research
- Evaluation of progress.

These changes require real commitment and coordinated approaches that are evaluated and embedded. We have already discussed the importance of coordinated women-in-engineering programs or similar initiatives in Chapter 3. In the following sections we examine some approaches, requirements and examples of change strategies which can be applied in the areas listed previously. At the conclusion of the chapter we focus on evaluation strategies.

In addition to internal policies and processes, external drivers can also play a role in fostering institutional change to promote inclusive education. In engineering education, external accreditation requirements can exert significant influence. Approaches to embed gender inclusive practices within accreditation requirements are therefore also discussed in this chapter. Another external driver which may promote change is government policy. In Australia, for example, since 1990 universities have been required to develop and report annually on equity plans for their institution. Since that year the Australian Government's Department of Education (under various names) has required institutions to report statistics on access, retention and success of six designated equity groups: Aboriginal and Torres Strait Islander people; people from non-English speaking backgrounds; people with disabilities; rural and isolated students; people from socio-economically disadvantaged backgrounds and women in non-traditional study (WINS). The non-traditional study fields for women were originally defined as those which had less than 40 per cent commencing domestic students who were female. In 1990 these were Agriculture; Architecture and Building; Business Administration and Economics; Engineering and Surveying; Science. From 2005 this measure was revised to include Engineering and Information Technology only since all other fields had achieved the 40 per cent target (the target for Engineering had been set at 20 per cent but this was not achieved) (DEST 2004). Equity plans have to report not only on the statistics but also on the programs and services that are being provided to improve the situation for each equity group. The initial result of this initiative was the removal of most of the obvious entry barriers that existed for such groups and the provision of access and support programs. However, progress has stalled in many ways since the reduction of government funding that was originally provided in support of the equity plan initiative.

Visible Institutional Leadership

In order to achieve institutional change, visible institutional leadership must be present at multiple levels, from top to bottom of an institution. As the case studies later in this chapter will demonstrate, some leadership starts from the bottom, from the faculty members who champion curriculum and cultural change and then encourage and support other faculty members to do so. However, even if faculty members develop support networks, or similar, to assist with this, they will ultimately need institutional support or risk foundering. This support may be in the form of funding for programs or administrative assistance but even more importantly higher level support must involve validation of their efforts, since, as observed in a recent National Academy of Science report: 'Those at the highest university levels must be convinced to take change seriously and give it high priority and high visibility' (NAS 2006: 8). This report provides some excellent 'Who can do what' checklists (NAS 2006: 114–117) that demonstrate a wide range of ways in which leaders at all levels of an institution can make a difference to its success in recruiting and advancing women students and faculty in engineering. The report contains a range of excellent suggestions for individual faculty members, Department Heads, Deans, Provosts and Presidents that make it clear that this is a responsibility of all levels of leadership. When high level leadership, at department and institution level, is seen to actively initiate and then continue to support the changes needed for inclusive education to occur, it is likely to be seriously addressed by all faculty.

One key area, unfortunately often ignored, where institutional leadership is required to support curriculum change relates to faculty workload and reward structures. Women faculty members are frequently allocated higher teaching loads in addition to taking on greater proportions of the student outreach and support roles, particularly, but not only, relating to women students (Bagilhole and Goode 1998). Attracting and retaining female students in engineering needs to be seen as an institutional responsibility, not one that is carried only by women faculty members. Institutions should also reward effective classroom innovations and practices along with the educational scholarship that should inevitably accompany such developments. These rewards should include opportunities for promotion and tenure (in countries where this concept is relevant) based on excellent performance in teaching and learning combined with scholarly research in this field. Ideally these measures would be accompanied by institutional commitment to effective evaluation of teaching and learning. Some institutions now, at least in theory, have such evaluation systems and promotion criteria in place, but there is often widespread cynicism from faculty members about the real commitment to such measures in practice. Faculty members who lead such initiatives and excel in the practice of teaching and learning, along with the related educational scholarship that accompanies it, must be seen to be rewarded in the same way as would a faculty member who excels in research in his or her technical discipline. Many engineering faculty members who are passionately committed to excellence in teaching and learning and who are the likely champions of developments such as gender inclusive curriculum, are currently pressured by Department Chairs and institutional structures to shelve that passion until they have secured tenure or promotion via the traditional research route. A very different message would be sent to faculty members if the Department Head co-authored a paper on a teaching and learning initiative with one of these faculty members instead, or if there were more Professors of Engineering Education. While many institutions frame their promotion criteria around Boyer's four areas of scholarship of the professoriate (Boyer 1990), there is a view that often little more than lip service is paid to Boyer's scholarship of teaching in faculty promotion and reward structures.

Equity/Diversity Policies and Practices

Some examples of policies and practices that are particularly relevant to increasing the attraction and retention of women in engineering education were explored in Chapter 3. These included the provision of flexible and alternative entry processes and bridging programs for students wishing to enter engineering without the required preparation in mathematics and science from high school, as well as support programs for women once they have entered engineering programs. But embedding equity requires more than this. It means that a wide range of people at all levels of the university take responsibility for equity within a wide range of contexts. Of course formal equity/diversity policies should be in place and should be seen to apply to all members of the university community. For example, faculty members teaching in a classroom situation should understand their responsibility under the university equity policy to ensure that no student is discriminated against or harassed on the basis of race, gender, religion and so on, as well as the procedures they should follow when such behaviour is observed. Department Heads should be required to report on equity issues in their annual plans and reports to senior management, and key performance indicators should be developed around this: for example, the percentage of female faculty members in engineering departments, and the percentage of these at professor level compared with men. University presidents should ensure that equity issues are included in setting the future directions of institutions and that these policies are adequately supported by funding for women in engineering programs, equity and diversity officers and the like.

Embedding Inclusive Culture and Curriculum Developments

Earlier chapters in this book have focussed on the theory behind, and practical guidance towards, the development of gender inclusive curriculum and cultures within engineering programs and departments. The focus here is on the means by which to ensure that such developments gain long-term institutional commitment. While there will always be a small but committed group of faculty who believe in such measures, who will pioneer their development and implementation and may also succeed in influencing and inspiring their fellow faculty members to follow their lead, this is not sufficient for long-term and embedded change. The disappointing reality is that too many faculty members are not very interested in improving their teaching, or they do not really care about poor student evaluations and hence will ignore any curriculum improvement measures unless they are compelled to consider them. If research outcomes (excluding education based research) are all that are seen to matter in institution promotion policies, funding decisions and public rhetoric, this situation appears likely to continue. This situation calls for institutional and departmental leadership, particularly in relation to evaluating and rewarding good teaching and learning, combined with hammering home the message that student evaluations of teaching do matter. In an inclusive curriculum, students should be credited with the capacity to make intelligent and serious comments on their teaching and learning experiences and poor evaluations should be addressed.

Apart from rewarding excellent teaching there are other measures that are important components in embedding curriculum change for engineering to become more gender inclusive. Of those which are entirely under the control of the university, the most effective are course and program approval processes and policies. If addressing inclusivity is a requisite in course and program statements and development processes, and attention to this must be demonstrated before a course or program will be approved, then this is a considerable step forward, although not sufficient on its own. A relevant example is discussed in one of the later case studies in this chapter.

Faculty Development

In any teaching institution, recommended improvements or developments in teaching, learning and assessment processes need to be communicated to the teaching faculty in persuasive and credible ways, rather than relying on faculty to read the education literature. In the case of gender inclusive engineering education it is likely that many faculty members will only have limited understanding about what it entails and how they could contribute to it. Professional development workshops in this area (and other areas of learning and teaching) will assist them to gain an understanding of what is involved as well as the learning theories on which gender inclusivity is based. Once again, such workshops need the visible and financial support of departmental and institutional leaders in order to encourage faculty members to attend. One example of such a faculty development initiative is the *Diversity Infusion Program* at Maricopa Community College in the US. Faculty members receive financial and mentoring support to attend a series of seminars over a year and utilize the resources and mentoring provided through the program to infuse diversity issues into a course that they are teaching in that year (Maricopa Community Colleges 2009). Faculty development opportunities do not have to be institution based. Relevant opportunities will be available in annual conferences of engineering education organizations, through collaborative projects between institutions and from external organizations. To get the most from such opportunities institutional leaders should sponsor not just individual faculty members to attend but also those at leadership level, such as Department Heads or Deans.

The most useful faculty development opportunities will involve much more than talk. Seymour (2002: 94) summarized the resources needed by faculty innovators to be able to proceed with curriculum innovations as:

- access to pedagogical and assessment expertise, preferably local to their institution
- teaching and learning assessment materials in accessible form
- digests of pedagogical and assessment techniques
- syntheses of the theoretical and research bases for these methods
- evidence of their efficacy, including reports of what has not worked well.

Some of these resources may also be available online through organizations such as *Diversity Web* (Association of American Colleges and Universities 2009).

Inter-Department and Cross-Institution Collaboration

Collaboration in many forms can assist with embedding gender inclusive engineering education initiatives and reform. Within a single institution, collaboration may take place between different engineering departments to develop joint courses that are gender inclusive. Some of these examples were discussed in Chapter 6. Inter-departmental collaboration may also occur between engineering departments and those in social science areas. Such collaborations can directly address issues of women in engineering through the development of courses that are partnerships between women's studies and engineering or science. Two such examples were described by Weasel et al. (2000) who considered modifications to an existing engineering course *Communications in the professional world* that included topics such as gender dynamics, communication styles and discussion of sexual harassment in the workplace. The second example they described was a new course for either women's studies or science students called *Gender and belief systems in physical science*. A German project, *Degendering Science*, has also developed a range of seminars in this area aimed at different target audiences and levels of study including future science teachers, gender studies students and students in science, mathematics and computer science (Bauer 2006). Collaborations of this type are often initiated by like-minded faculty in different departments, but they will require support from higher level leadership to be sustained. In other cases such collaboration may be imposed from above, and these cases will require time and effort to be spent on faculty development and encouragement to be successful.

Other collaborations are between institutions. Two such case studies are included in this chapter. Many of these collaborations are encouraged by national funding bodies, or international organizations such as the European Union, which provide major research and teaching grants to support them. Such collaborations have many advantages in terms of spreading the workload involved in any major change initiative, demonstrating significant national and institutional support for initiatives and hence meeting the need for visible leadership and ensuring wide dissemination across institutional boundaries.

Ongoing Research

In order to evaluate the effectiveness of curricular innovations and interventions as well as other initiatives such as those described in Chapter 3, it is essential that ongoing research in these areas is conducted and supported. However, there is still a lack of widespread published research available in this area. Often the hard-pressed faculty members (frequently women) who support much of this type of work do not have time to thoroughly research and then publish it, and when they do, such publications may not be counted towards tenure or promotion, or at least not weighted as heavily as traditional discipline based research in their field. In addition, if such research is not published in mainstream engineering journals, then other faculty members who might be influenced by it will probably never see it (Cronin and Roger 1999). Research by Foertsch et al. (1997) cited by Seymour (2002) also indicated that positive findings reported from research on curriculum interventions and developments, or even demonstrated through video evidence, may be ignored unless the institution and the presenter have high research prestige. Other reasons that can contribute to the lack of research on program and curriculum effectiveness are the lack of relevant statistical data, lack of resources, lack of understanding or acceptance by engineering colleagues of education evaluation research methods and results and the fear that poor results might result in discontinuation of a program or intervention, rather than encourage further work to improve it (Cronin and Roger 1999). For gender inclusive engineering education to achieve widespread acceptance and implementation, national and institutional support for ongoing research into its effectiveness is essential.

CASE STUDIES

Three case studies are provided here as examples of efforts to embed gender inclusive engineering education initiatives into institutional policies, practices and consciences. Two of them were not explicitly gender focussed but fit the criteria we have described earlier for gender inclusivity. In the first case at the University of South Australia, inclusive curriculum measures in engineering formed a subset of an institution-wide project. The Gateway Coalition in the second case study was a multi-institution collaboration that focussed on the development of integrated engineering curricula that resulted in improved gender inclusive outcomes. The third case, the ADVANCE project, focussed on issues relating to female engineering faculty, another important component of gender inclusive engineering education discussed in Chapter 3.

Gender Inclusive Engineering Curriculum at the University of South Australia

The University of South Australia (UniSA) was established in 1991 through the merger of several antecedent institutions including an Institute of Technology and former teacher training colleges. The University's founding act and subsequent mission has always stated that it should provide education to a range of people from educationally, socially and culturally disadvantaged backgrounds, and hence it has a range of access pathways and programs to support this (University of South Australia 2008b). The most recent statistics indicate that 44 per cent of its domestic undergraduate students fall into one or more of the recognized national equity groups. Civil, mechanical and electrical engineering programs are offered by three separate departments in one division. The percentage of female engineering students averages around 12 per cent, slightly low in comparison with other Australian universities and probably explained by UniSA offering none of the engineering programs that have been shown statistically to be more attractive to women students, such as environmental, chemical, biomedical or industrial engineering. Equity statistics available for 2008 indicated that 26.7 per cent of UniSA's engineering students were identified as low socio-economic status according to Australian government definitions, slightly higher than the university's overall percentage of 25.4 per cent (University of South Australia 1996–2009).

In 1997–1998 UniSA conducted an 18-month inclusive curriculum project across all programs in the university. The project aimed to develop inclusive curricula by improving the understanding and practice of faculty and developing guidelines to assist them in restructuring their courses to become more inclusive. The project was intended to raise awareness of the issues; influence institutional and departmental policy, by providing faculty development; and develop and collect resources to assist the growth and extension of inclusive curricula after the formal project.

Visible leadership for the project was provided by the university's Vice Chancellor (or President, in US terms) in initiating the project and then through speeches and policy statements supporting it. The project was co-directed by the Pro-Vice Chancellor (Equity and Academic support) who also chaired its two key committees, thus emphasizing the commitment of senior management to the project as a whole. The support of the engineering departmental heads was particularly strong during the project, and this led to the provision of additional drive to develop inclusive curricula within those departments.

Following an audit to assess the extent of inclusive curriculum understanding and practice amongst faculty, workshops were developed on topics such as 'Towards an Inclusive Curriculum', 'Developing an Inclusive Curriculum' and 'Learning Styles and Their Implications for Students' Success'. Attendance at workshops was not compulsory but was strongly encouraged by most Heads of Departments, particularly in engineering. As expected, most attendees were those faculty who were already identified as having a strong interest in teaching, but this included the vast majority of those who taught large, early year engineering courses and approximately 30 per cent of engineering faculty participated in one or more of the workshops. A website was developed along with an internal manual titled *Making Engineering More Inclusive* that was circulated to faculty members (and was the genesis of this book).

At the conclusion of the project the principles of developing inclusive curriculum were formally incorporated into two university documents: the *Policy for Development, Amendment and Approval of Programs and Courses* and the *Code of Good Practice: University Teaching* and thus were mainstreamed within the university. It became a requirement that when planning the development or amendment of programs, faculty should:

> . . . indicate in what way advice has been sought on issues of inclusivity, how that advice is acted upon in the planning and delivery of the program, and what mechanisms will be adopted to evaluate the level of success of such program components.
>
> (University of South Australia 2000)

The project was funded for an 18-month period and clearly within that time frame it would be unlikely that major impacts of the faculty education process could be evaluated in terms of improved access, success or retention figures for female students. It is now over 10 years since the formal, funded inclusive curriculum project took place and, as is probably the case with many universities, the University of South Australia has undergone continual change since then. With respect to engineering and the way in which it is taught and administered, some of these changes have been structural and some have come full circle. In 1997 when the original project commenced, all engineering programs and all engineering students and faculty members belonged to a single Department of Engineering and the first year of the program was common to all engineering students. By the time the funded project was completed the single department had been split into three, and each gradually moved to redevelop much of its program curriculum over the next 10 years. Although the three department structure still exists, the common first year was suddenly reintroduced in 2008 and has led to many more curriculum changes. Maintaining curriculum advances such as those achieved during the original inclusive curriculum project could therefore be considered to be problematic in such an environment.

Throughout these changes the policy embedding achieved in the original project has been largely maintained. Most notably, the amendments to the approval processes for the development of new programs and courses that require the consideration of inclusivity have been maintained, although the requirement for evaluation has been removed. The equivalent 2008 guidelines read:

> Identify how the curriculum of the program has been developed to take account of the perspectives of non-dominant groups. This should include consultation with staff of the University. See ideas and examples of educational strategies for improving inclusivity in curricula, teaching, learning and assessment at . . .
>
> (University of South Australia 2008a)

The document links to a university web resource on inclusivity which includes the guidelines for educational strategies for improving inclusivity in curricula, teaching, learning and assessment that were developed during the original project.

As mentioned earlier, Australian universities are required to report annually to the federal government on the access, participation, retention and success rates for defined student equity groups, based on the enrolments of domestic students. (These statistics do not include international students.) One of these equity groups is Women in Non-Traditional Studies (WINS), which consists of women in engineering and information technology. In 1996 (the year before the inclusive curriculum project commenced), access statistics for this equity group were 12.5 per cent at UniSA compared with the national average of 14.6 per cent. In 2000 the UniSA figure had increased slightly to 12.6 per cent and the national average to 15.2 per cent. But by 2005 both these figures had decreased: the UniSA percentage had slipped to 12.2 per cent, but the national figure had slipped by a greater amount to 14.1 per cent (DEST 2004; University of South Australia 1996–2009). UniSA retention data in 2006 were basically the same for female and male students in engineering at 78 per cent, slightly below the overall university average of 82 per cent. However these figures are significantly higher than the estimated national averages for engineering in Australia of 58 per cent retention for women and 51 per cent for men (King 2008). It should be pointed out that it is not possible to attribute any improvement or success in these statistics solely to curriculum improvements that encompass inclusivity (or vice versa), as many other factors and initiatives are involved, both internal and external to the departments.

Perhaps the most significant feature to emerge from the UniSA experience is that gender inclusivity has to be thought of as a continuous process rather than as a feature that can be dealt with once and for all by the incorporation of appropriate curriculum content. Gender inclusivity involves continual reflection on the teaching and learning interaction and a constant alertness to the potential of the style and content of teaching to create situations where some students may not feel included and consequently may choose to exclude themselves from further study.

Drexel University and the Gateway Engineering Education Coalition

Drexel University is a private institution established in 1891 in Philadelphia, US, and is now the third largest private engineering college in that country. It offers a full range of traditional engineering programs as well as those known to attract higher female enrolments such as environmental, chemical, biomedical and architectural engineering (Drexel University 2009). In 1987 Drexel University instigated a major reform to its undergraduate engineering curriculum. A committee was set up to examine the entire framework of the undergraduate program. It began this task by developing a set of characteristics that 'future engineering graduates should possess to become leaders of the profession' (Fromm 2003: 113). In the first year of the development 45 faculty members participated in a three-day workshop to explore their current curriculum practice in the separate departments and to work towards comprehensively restructuring the curriculum.

The curriculum restructure that was undertaken was focussed on integrating courses in the freshman and sophomore years using four interwoven sequences, namely, mathematical and scientific foundations of engineering; the art of engineering; energy, systems and materials; and personal and professional enrichment program (Fromm 2003). The focus was on students' early involvement in engineering, bringing mathematics and science into context as underpinning these and on emphasizing engineering design, interdisciplinarity, experimentation and communication skills as core elements of engineering. Students were required to complete laboratory projects totalling four hours per week in each of the first five terms and to complete 10 design project experiences in the first two years, compared with none in the previous curriculum. In 1998 a National Science Foundation (NSF) award was obtained to sponsor the curriculum development under a project titled *Enhanced Educational Experience for Engineers* (E^4) which enabled expansion of the program and facilities required for it. Faculty members involved in the development came from 13 different departments including humanities, arts, science and engineering. The revised curriculum was introduced as an experimental program initially, with students given a choice to undertake it or remain in the traditional curriculum. The first students to enrol in the experimental curriculum commenced in 1990 when 100 of the 600 students elected to do so (Froyd and Ohland 2005). After the first two years of the experimental curriculum, retention rates for those students averaged 21 per cent higher than students in the traditional program (Fromm 2003). This early success led to the replacement of the traditional curriculum by the integrated one for all students from 1994.

Due to the demonstrated early success of the integrated curriculum, other universities became interested in doing something similar and in 1992 the Gateway Engineering Education Coalition was formed and was subsequently supported by two NSF grants. The Gateway Coalition consists of seven institutions of varying backgrounds and locations across the US. The original aim of the Coalition was to extend the Drexel freshman/

sophomore model to all of the institutions and then to expand that to the full four-year program. The Gateway Coalition also has a specific focus on increasing the percentage of underrepresented minorities, including women, to graduate from their programs and has therefore monitored statistics relevant to this from the start. The Gateway Coalition recognized the need for visible institutional leadership in order to achieve commitment from faculty across a range of institutions rather than the single initial institution at Drexel. Hence the management structure of the Coalition consists of a Council of Deans, to focus on policy issues that need to be addressed in their individual institutions; a Governing Board, to establish the direction and agenda and provide the intra- and inter-institutional connections; and an External Advisory Board (Fromm 2003).

For a full discussion on the process and range of curricular and faculty development undertaken by the Gateway Coalition, the reader is referred to Fromm (2003) and the Coalition website (Gateway Engineering Education Coalition 2008). The major focus of the Gateway Coalition has been on the development of integrated curriculum, but the Coalition has also sponsored a wide range of projects related to access and retention for female and other minority students, as well as a range of faculty development initiatives. We will focus here on the success of the program, particularly for women and minorities. Between 1992 and 2002 statistics were gathered on a range of indicators of student and faculty involvement and student success and retention. The Coalition was successful in significantly increasing the numbers of students involved with engineering design, the faculty involved with teaching the courses, the number of credit hours of integrated curriculum courses offered, the number of higher year courses taught by interdisciplinary faculty teams, the number of faculty attending engineering education conferences and several other measures (Gateway Engineering Education Coalition 2008). The most striking of these achievements in relation to gender was the significant improvement in retention rates when compared with science, mathematics, engineering and technology (SMET) focussed programs in other institutions in the US. By the fall of 2002 the retention rate for female students in engineering from first to second years across the Gateway Coalition programs was 88 per cent (compared with 83 per cent in 1992 when the Coalition began). This compared with a rate of 68 per cent for SMET schools nationally. Retention rates for women from second to third year were also maintained at 88 per cent. Between 1992 and 2001 the percentage of engineering degrees awarded to women at the Coalition schools increased from 13.6 per cent to 19.5 per cent. All of these are very clear endorsements for the integrated curriculum approach that was adopted and indicate that such an approach is clearly gender inclusive. Retention and graduation rates for African-American students showed similar or even greater improvement and, significantly, the rates for the student body overall also improved. This outcome endorses our earlier assertions that measures introduced for gender inclusivity are in fact inclusive of, and successful for, all students, not just women.

Although the Gateway Coalition is no longer funded by the NSF many of the curriculum innovations that were introduced are still operational today and their website provides a rich array of resources that are still useful to engineering educators wishing to implement gender inclusive curriculum innovations (Gateway Engineering Education Coalition 2008).

ADVANCE: Institutional Transformation program

The ADVANCE initiative has been funded by the NSF in the US since 2001. It is a competitive program that awards Institutional Transformation Awards of up to $750,000 per year for up to five years to promote the increased participation and advancement of women in the science and engineering fields (Rosser 2008). The primary focus of the grants is to advance the careers of women faculty members by enabling them to move into senior ranks and leadership positions. As we noted in Chapters 2 and 3, women remain severely under-represented in faculty positions in engineering, and even more so at the senior levels of Professor. Women faculty members provide important visible role models to undergraduate and graduate female engineering students, and hence increasing their representation is another important component of gender inclusive engineering education. Previous NSF funded programs such as the Professional Opportunities for Women in Research and Education (POWRE) had focussed on support to early career women to enable them to gain tenure and promotion. However, when the POWRE program was evaluated it was found that in some cases the opposite effect had occurred, as the grants obtained by women under that scheme were assumed not to be competitive and hence were devalued by promotion committees (Frehill 2007). As a result the ADVANCE program was very specifically aimed at institutional transformation rather than the previous focus on 'fixing' individual women. In this regard it provides another illustration of the progress in thinking that has occurred in support initiatives for women in engineering as discussed in Chapter 3.

The first nine ADVANCE awards were made in 2001–2002, with 10 further grants made in 2003 and two further rounds since then. Frehill (2007) described the range of programs that were implemented as a result of the first two rounds of grants and analyzed their success. Of the first two rounds of grants, 15 went to institutions that had engineering programs. These 15 engineering schools were quite diverse in terms of overall size, existing percentage of women faculty in engineering, research expenditure and the proportion of overall degrees awarded by the institution that came from the engineering area (Frehill 2007). Although the ADVANCE awards are focussed on specific institutions, these universities also engage in a range of collaborative efforts to share ideas and experiences in order to adjust their individual programs and have a common web portal which provides access to a wide range of material relating to the implemented programs (ADVANCE 2009).

Rosser (2008) has described the five major components of the ADVANCE program at Georgia Tech, which received funding in the first round. Georgia Tech is one of the largest engineering schools in the US with over 400 faculty and over half of all degrees awarded at the institution being in engineering. These components were:

- four tenured female professor positions (one of which was in engineering) funded for five years and charged with the responsibility to form a mentoring network for the women faculty in her college
- collection of gender disaggregated data on faculty appointments, rank, funding, space and so on
- family-friendly policies and practices
- mini retreats to facilitate access to those in decision making positions and allow informal discussion on topics important to women faculty
- removal of gender, racial and other biases in promotion and tenure processes.

The strategies implemented at Georgia Tech were typical of many of the ADVANCE programs at other institutions, but each program was tailored to the particular culture and situation of the individual institution. Frehill (2007) describes a range of additional strategies that were also implemented in the areas of recruitment, retention and advancement at various institutions. These included start-up funds for newly hired female faculty, dual career couples policies and faculty and leadership development programs. In analysing the impact of ADVANCE on engineering programs Frehill (2007) notes two important factors that we describe as:

- where the program was seen as a 'woman's program', male faculty did not participate, even though the activities may have been available to them
- it was imperative to work with the engineering leaders (usually the Deans and Department Chairs) to implement the program and for them to be involved at the planning stage.

As Rosser (2008) notes, the challenge for those institutions which received ADVANCE grants and implemented a range of programs as a result will be to sustain those programs when the funding from NSF is no longer provided after the five-year grant is ended. She endorses the importance of the factors that we noted earlier in this chapter being in place to enable this to occur, namely, leadership at institutional, administrative and faculty levels and penetration of the structural, procedural and cultural levels of the institution.

EMBEDDING CHANGE IN ACCREDITATION PROCESSES

Engineering programs in most countries are subject to accreditation by the relevant national professional engineering body in that country. Details of a range of accreditation criteria are provided in Appendix A. Accreditation bodies have considerable power and influence over the curriculum and the leverage to change it. Historically, accreditation criteria were often cited as a reason to maintain the status quo, and many proposed curriculum initiatives foundered on the accreditation rock. However, more recently, accreditation boards have begun to promote change in the undergraduate engineering curriculum and pedagogy by becoming much less prescriptive about absolute requirements for content and structure in favour of specifying outcomes that programs are required to achieve. This has opened the way for significant curriculum change in some institutions. But can accreditation criteria also exert influence towards increasing the gender inclusivity of engineering curricula?

Currently neither the US ABET accreditation system nor the European EUR-ACE process has any explicit or implicit direction towards gender inclusive engineering education. In fact, ABET has stated that 'it should not mandate diversity goals as part of its standards' (ABET 2005). However, in Australia the accreditation board of Engineers Australia has made some significant moves in this direction in its most recent accreditation process update. Engineering programs are assessed on three criteria: the operating environment, the academic program and quality systems. The expectations for each of these assessment criteria are interpreted through a range of performance indicators which are intended to be used as guidelines rather than prescriptions. Within the performance indicators there are now explicit references to gender inclusive cultures and curriculum. For example 'gender balance' in academic and support staff profiles is one indicator, thus implying that universities should increase the numbers of female faculty to work towards this (since these numbers are low in Australia as they are worldwide, as described in Chapter 2). Another performance indicator relating to the academic program is 'Inclusive environment—gender, culture, social differences—encouraging diversity and the development of the individual', and another is 'flexible structure adaptable to student backgrounds and individual learning abilities' (Engineers Australia 2008a). These developments had their foundations in the major review of engineering education in Australia that was conducted in 1996 (IEAust 1996a) and the subsequent review of progress towards its recommendations in 2008 (King 2008). By making issues of gender and inclusivity explicit in their accreditation processes, Engineers Australia is expressing its commitment to the goals of gender inclusive engineering education and requiring universities to pay attention to it.

EVALUATING CHANGE AND SUCCESS

When access and retention initiatives or curriculum changes are implemented, it is essential that their success (or otherwise) be evaluated in order to demonstrate to those involved that their efforts are acknowledged as valuable and worthwhile as well to gain institutional support for their continuation. Three important components of such evaluations are now considered: data, benchmarking and meta-analysis.

Data and Statistics

The maxim 'What gets measured is what gets done' is widely known in business. If something is not measured it is assumed that it is not important. Without data it is also very difficult to convince others (especially engineers) that something is important and needs addressing. Many people are actually unaware that the numerical situation for women in engineering is still so poor: they assume that since universities have as many, if not more, women than men that this must hold true for all programs. Thus Cronin and Roger's assertion is important: 'We believe strongly that gathering, presenting, and articulating this information are essential in establishing the rationale for concerted action to redress this underrepresentation' (Cronin and Roger 1999: 639). For this reason many studies and government reports have recommended that comprehensive data, disaggregated to the level of equity groups, must be collected within institutions, workplaces and across nations in order to demonstrate that issues do exist for women and minority groups in engineering and then to monitor any progress that has been made. In the US, data on women, minorities and persons with disabilities in science and engineering have been collected by the NSF since 1984. In the European Union the ETAN report (2000) strongly recommended that organizations should collect reliable gender-disaggregated statistical data in order to monitor and review the effectiveness of policies and that such data should be disseminated in 'user-friendly' formats. The SET Fair report in the UK (Greenfield et al. 2002) resulted in the establishment of the UK Resource Centre for Women in Science, Engineering and Technology (2009) which now provides an excellent range of statistical data in this area.

Whilst there is a need to collect such data on a broad scale, there is also a specific need to collect data to assess the effectiveness of any gender inclusive engineering education measures, whether they are attraction and retention programs or curriculum changes. Statistics alone are not sufficient. Qualitative data through surveys, interviews and the like are also necessary to elaborate on the reasons behind student choices and experiences, and to gauge reactions to curriculum or program interventions and initiatives. Gathering such data may also require collaboration with experts from outside engineering departments, such as colleagues from education and the social sciences. Recommendations from the *Women's Experiences in College Engineering* report (Goodman Research Group 2002) on data collection and analysis approaches for this purpose include:

- brief questionnaires to engineering students at periodic intervals
- longitudinally tracking individual students in engineering departments from entry to exit
- integrating data keeping with monitoring/mentoring of all students.

These data must also be collected in such a way that disaggregated results for women and other minorities can be determined.

Benchmarking

How does your gender inclusive practice compare with that of others? How do students and faculty members perceive the interventions being trialled in your course compared with other courses? Are retention statistics for women in engineering at your institution better or worse than other comparable institutions? These and other such questions can be at least partially answered by benchmarking. We are not suggesting that benchmark figures should be used as absolutes but rather to indicate trends and areas that need further attention.

A range of different benchmarking tools have been developed for issues relevant to women and other minority groups in engineering education. One example is the *Diversity Scorecard* (Bensimon 2004), now renamed the *Equity Scorecard,* which was developed at the University of Southern California in 2001 and has subsequently been used at over 40 colleges in the US (CUE 2009). Although this measure was not specifically targeted towards women in engineering, it could be readily adapted for this purpose. The scorecard involved a three stage process. Firstly basic data were disaggregated by race and ethnicity on four perspectives: 'access, retention, institutional receptivity, and excellence' (Bensimon 2004: 46). On the basis of this information performance goals were then identified for each of the perspectives. The final step was reporting to the Presidents of the respective institutions, emphasizing again the need for leadership response to address these issues.

The University of Newcastle in Australia has developed a benchmarking resource specifically to assess cultural change in engineering education (Jost 2004). Subsequently we have adapted parts of this resource to develop a questionnaire to assess the perceptions of engineering faculty members and students about the inclusivity of their courses and programs (Mills, Ayre et al. 2008). The questionnaire presents statements about various aspects of the curriculum, and respondents are asked to indicate their level of agreement with the statements on a Likert scale, in relation to the courses they teach or the degree program that they are studying. A sample question is provided below:

Question 8: How is theory taught within your program?
- Level 1—Theory is taught largely in isolation
- Level 2
- Level 3—Theory is presented in terms of specific industry related problems
- Level 4
- Level 5—Social effects are considered and debated wherever possible.

To date we have used the questionnaire in engineering departments at four institutions in Australia, one in the US and one in UK. Maybe not surprisingly, at all of these institutions, faculty generally perceived that their practice in courses they teach was more inclusive than students perceived to be the case for the overall program they were studying. The greatest perception gaps between faculty and students varied slightly between institutions but generally related to how theory was taught, the type of problems used and their multi-disciplinarity. One pleasing result has been that students perceived that issues of discrimination, harassment and unacceptable language were addressed strongly by their institutions and indicated that the benchmark level was between four and five in each case (with five being excellent performance) (Mills, Ayre et al. 2008).

Meta-Analysis

The term 'meta-analysis' refers to a study which examines the outcomes of a range of experiments or published results in the same subject area and analyzes them to establish overall trends and make recommendations. We have already mentioned several such studies in earlier chapters (for example Thom 2001; Froyd and Ohland 2005; Heywood 2005; Phipps 2008). One other example is the *Building Engineering and Science Talent (BEST)* program which was a public-private partnership that produced three reports in 2004 that examined best practice for women and minorities in science at engineering in the K–12, higher education and professional workforce sectors (BEST 2007). All of these meta-analyzes can provide valuable insights and information that can be used to assist the initiation and evaluation of new curriculum and other initiatives to increase access and retention of women in engineering education.

SUMMARY

This chapter has examined some of the policies, strategies, processes and systems needed to embed gender inclusivity into an institution, and illustrated these through three institutional and cross-institutional case studies. Critical among these is visible and involved leadership, combined with appropriate faculty development to develop inclusive curriculum and departmental cultures. We have also considered the possibility of embedding gender inclusive practice into national professional accreditation processes and policies and provided one example where this has come close to fruition. Finally we have discussed the importance of evaluation as a means of gaining the evidence needed to improve and embed gender inclusive policies and practices and offered some examples of available instruments that could be used to do this.

8 In Conclusion
A Look to the Future

INTRODUCTION

This chapter provides an overview of the general argument of the book and demonstrates the logic of its development by reprising the themes set out in the sequence of chapters. It concludes with some reflections on the size of the task involved in making engineering education more gender inclusive and urges its message as key to transforming engineering education in this century.

WE STILL NEED MORE ENGINEERS!

We began this book with an overview of the many arguments that have been presented in the popular press and in the engineering education literature which have re-iterated the message of the need for more trained engineers. While some elements of this shortage are connected with particular locations and functions, or have become acute at particular times, the breadth of the call for more engineers presents the situation as a global crisis. Without more trained engineers many of the innovations needed to cope with current global issues, such as global warming, environmental change, automobile modifications, more and faster global communications and more resource exploration, to name but a few of the most urgent, will be unavailable to future generations. The world needs more engineers and it needs them now.

Global movements and recognition of the value of credentials in the current knowledge economy have led to a rapidly increasing diversity of the student body in many higher education institutions. A qualification in engineering is understood as desirable in global terms as it has clear value for the progress of society as well as for the individual. Not surprisingly then, engineering is one area marked by increasing numbers of international students in many higher education institutions. However the advent of students from different backgrounds into engineering classrooms around the world has not often been accompanied by related changes in traditional teaching and learning practices or assessment strategies. When these non-traditional students have trouble achieving the desired standard, they are sometimes identified as being under-prepared and not able to cope with the demands of the course. In the previous chapters we have variously suggested that in fact any lack of achievement by these students may be ascribed to their learning needs not being met, their approaches unrecognized and their values dismissed. We have urged engineering educators to become more aware of their teaching in order to be able to adapt to learning needs which differ from those of the traditional engineering student. However there is frequently an ingrained resistance to being able to see the student's struggles as the product of their particular situation rather than as a simple indication of 'not being up to it'. Only recently are some educators responding to those who dismiss non-traditional students as being less than adequately prepared for their courses. As noted in a recent edition of the 'Tomorrow's professor' online newsletter:

> Non-traditional students are not underprepared. Their preparation responds to a different way of seeing themselves and understanding the world that derives from their own cultures and traditions. This different way of seeing the world has repercussions in most academic areas. They influence the way students think, express themselves, interact in the classroom, and think in the disciplines. For example, many non-traditional students tend to see things in a subjective, inward-looking fashion (Haigh 2009). Other students from non-Western societies are holistic in their thoughts. They tend to emphasize and value how things are interconnected. They tend to give contextual and emotional information.
>
> (Hermida 2009)

As we have demonstrated in earlier chapters, an inclusive style of education would involve a good deal of sharing of different perspectives. One outcome of this process is to make learners much more aware of their own thinking, especially in the ways it differs from that of their peers. Consequently we have recommended a constructivist approach to teaching and learning in which both students and instructors are encouraged to become more conscious of their own styles of thinking and learning. There are clear responsibilities for the teacher in this scenario. For instance, the person leading the group or the class would ideally make a clear statement about what is expected in terms of classroom involvement, opportunities for further discussion (not limited to 'If you have a problem you can come and see me'), behavioural expectations, along with the usual material about assessment criteria and course structures.

In many respects the excerpt from Hermida (2009) quoted earlier could apply equally well to women students, the particular focus of our work in this book. As we have shown earlier, female students may bring different sets of expectations and learning needs into the engineering classroom. Alternatively, some of them may be just as focussed on the usual engineering applications and as eager to learn in the standard way as are the men. Others, and particularly those from different language and cultural backgrounds, may need clearer specification of the expected practices in order to be recognized as part of the group of learners. What we have suggested throughout this work is that it is incumbent on university faculty to devote some time and energy to finding out about student expectations and orientations, to develop a sense of mutual engagement in knowledge production and to eschew the position of the 'one-who-knows' whose job it is to fill up the empty heads.

THE CURRENT SITUATION OF WOMEN IN ENGINEERING

We showed in Chapter 2 that although the under-representation of women in engineering education has been a constant feature of the higher education sector across the developed world, it was not recognized as a problem until the late twentieth century when, as a consequence in part of the rise of the women's movement, attention was paid to the position of girls and women in educational systems. While the minority position of women students in tertiary education generally has been reversed in the past 30 years, especially since education and nursing programs became part of the university sector, engineering has continued to be the province of a clear male majority. Although there are indications that in some cultures women are more widely recognized as engineers and the numbers of entering female students are closer to those of the men, the dominant pattern of engineering as a male educational stronghold has continued.

STRATEGIES FOR INCREASING THE ACCESS AND RETENTION OF WOMEN INTO ENGINEERING STUDY

The past two decades have seen a considerable number of strategies aimed at improving women's enrolment numbers and experience in order to assist them to begin, and then to complete, engineering degrees. Some of these interventions have begun in schoolrooms as early as kindergarten level while others have directly addressed the enrolled students on campus. Many of these strategies have produced demonstrable gains for women in terms of numbers and completions. However, some of these gains can come at a cost, particularly in terms of time and energy of the female faculty usually involved, many of whom are already disadvantaged by being a minority within their faculty. This is not to say that only women professors have addressed the on-campus issues: many men have also joined the campaign to encourage more women engineers. However, the majority of work in this direction has been done by women on top of their regular teaching, research and administrative work.

While in some places affirmative action strategies have been deployed and the idea of quotas in terms of an appropriate number of women to be enrolled has been discussed, these measures can be associated with backlash. The most terrible example of this was the shooting of 14 women engineering students in a Canadian university in 1989 by a male who had been rejected for entry to engineering on the basis of insufficient grades and blamed the admission of women students for this. Other, less dramatic examples are more common with some women students feeling they are regarded as getting an easier run or treated more leniently because of being women. These women want to be seen as competent and good students in their own right, and they reject the idea of needing special treatment.

As was shown in Chapter 3, resourcing is also a factor. When the money for a particular program runs out or when a key person leaves, the program is often discontinued. Not surprisingly, these strategies have produced some clear gains but not long-standing or ongoing ones. And so the situation of women being under-represented in engineering enrolments continues across the board.

THE MISSING FACTOR

In this book we have pointed out, along with some other recent voices (Busch-Vishniac and Jarosz 2004; Pawley 2004; Riley 2008), that few of the strategies to improve the lot of women and minorities in engineering education have directly addressed issues of curriculum. In fact it seems that many engineering educators believe that the curriculum is sacrosanct, comprising as it does the key knowledge requirement for engineering work. Our own studies have encountered many examples of faculty resistance to thinking about gender at the same time as thinking about curriculum, especially in terms of content, as exemplified by the following faculty responses to the benchmarking surveys we conducted in 2009 and have described in Chapter 7:

> I would expect that a modern approach to curriculum development and delivery would be completely agnostic to gender of the student body. I can't see why gender should be a consideration in any of the technologies, concepts, learning processes, etc.—students should be considered 'students' and efforts and energies should be directed at ensuring that the students as a whole are encouraged to participate, contribute and enjoy their learning—gender shouldn't be a distraction.
>
> (Faculty member, Male)

> I do not have any specific examples of gender inclusive practice in the subjects that I am involved with. In most of my subjects, the content

> is mostly technical and, as far as I am concerned, each student's ability and performance in the subjects and each student's understanding of and competence in the subject material is unrelated to his/her gender.
>
> (Faculty member, Male)

Unfortunately the line of thinking that sees gender as a 'distractor' bespeaks a lack of awareness of the need to think about the people in the classroom rather than the content of the subject to be learned. Engineering educators who take refuge in the idea that their knowledge is 'purely technical' do not completely understand the learning processes in which they are engaged. Everyday life experiences attest to the fact that very few learning encounters are 'purely technical'. Any household purchase, such as a car, a dishwasher, a telephone or sound system, often involves a salesperson with a high level of technical expertise. But the purchase is also coloured by the degree to which the salesperson is good at explaining the machine, can respond to questions, can engage with the needs of the buyer and emerges as personally involved in the process. In a similar way the educator who is clearly involved in the material, keen to ensure that the students grasp the knowledge, open to discussion and open to multiple approaches to problems is much more likely to be effective than one who sticks to the book and displays no personal involvement at all, irrespective of the nature of the content.

One of the issues here is that keeping the content separate from the students' interests and concerns may have been the traditional approach in a good deal of university education. This pertains to the older term 'syllabus': the list of essential knowledge and skills that must be covered in any course. Courses were, and in formal course documentation often still are, identified in terms of their content which was seen as knowledge to be given out. Whether or not the students learned the knowledge and succeeded in the course was explained in terms of their ability or lack of it. We have argued that in fact such an approach is impractical in current engineering classrooms if they are to respond to the increasing student diversity in ways that encourage and promote learning for all students. Like many educators, we use the term 'curriculum' as embracing not only curriculum content, or what is to be learned, but also pedagogy: how it is to be taught and to whom, factors about the students, their prior learning and their expectations. We argue that it makes no sense to speak of one without attention to the other, that content and styles of teaching are implicitly interlinked in any evaluation of the experience of university teaching. The composition of the student group is the other key feature of our attention to inclusivity. Hence we urge engineering educators to attend to the gender dimensions of their classrooms in order to ensure the women are as equally involved in the teaching and learning process as the men.

STUDENT VIEWS ON THE GENDERED CULTURE OF ENGINEERING

Many faculty members and students believe that issues of gender do not arise in the engineering classroom environment at universities any more: that those days are long gone. And yet, student responses to our recent benchmarking survey still indicate otherwise. When students were asked about the degree to which their learning experiences in engineering had included inappropriate gendered behaviour and references, some fairly traditional sexist comments still arose:

> If the engineering/construction industry is male-dominated, there must be some reasons for it, e.g. men are generally more logical, rational and effective at solving technical problems.
>
> (Student, Second Year, Male)

> Lecturers give special treatment to females to ensure (such as alternative exams when they fail a subject) so they do not withdraw from the course of engineering but not for gender equality purposes but for marketing and promotions to ensure the university maintains its quota.
>
> (Student, Fifth Year, Male)

Both of the male students cited display sexism and cynicism in ways antithetical to the promotion of gender equity. Their experience clearly involves more than 'purely technical' content insofar as their gendered attitudes are amplified by their education, rather than challenged. (And, of course, the university does not have a 'quota' to fill for female students).

Some women students also tend to reject the idea that inappropriate comments were a problem, as in the following comment from a woman student who had decided to quit the course:

> The lecturers do sometimes comment in inappropriate ways but I personally am not offended (just put out) and recognise that it builds rapport with the rest of the class and relevance. You have to be tough to succeed in this industry and I am not tough enough.
>
> (Student, Third Year, Female)

This speaker has adopted the 'one of the boys' orientation and resists the idea of being offended but at the same time has made a decision to leave as she feels she is not tough enough. Her stance is an example of the conflicting positions around gender issues that emerge in the engineering education research.

While these comments show that both faculty members and students reject the idea that the engineering classroom climate is permeated with the sort of masculine style that is commonly found there, there are others who do recognize its gender bias, although no one is held responsible. An example of this, from a female student in the same survey, is as follows:

> It's tough working and going to class with 100 guys who all think they are alpha males who are totally awesome. Sometimes it feels like a massive . . . contest in class. Not the lecturer's fault, and not many women have stuck the course out and that says a lot about how this is dealt with.
>
> (Student, Fourth Year, Female)

This speaker clearly recognizes the male dominance of the classroom style and can see how women find it difficult. However she draws back from blaming anyone by saying that it is 'not the lecturer's fault'. We have argued in this book that the person in charge has a responsibility for classroom management and that there should be some basic rules about which all agree, that centre around the idea of mutual respect. Given the indications from this current research it would appear that it will be some time yet before the message is taken up by the entire engineering education community. And yet the need for this work has never been more pressing. The engineering profession cannot afford to maintain sexist attitudes and processes that are significantly at odds with broad community understanding. To do so would be to maintain the dearth of women engineers along with that of other non-traditional enrolments whose presence among qualified engineers is so keenly anticipated.

LEARNING PRINCIPLES

In Chapter 4 we outlined the principles of constructivist approaches to learning which provide a model of personal and professional engagement with the material to be learned and with the other learners. Understanding learning in this way changes the classroom from an arena for the transmission of information to one of a community of practice in which students and faculty are mutually involved in the learning process. The benefits to be gained include broader recognition of the range of approaches to learning typically found in any classroom. As we have maintained, this approach sets out particular forms of responsibility for the teachers along with the requirement to become increasingly reflective about their own practice. The traditional blaming the victim approach, 'I taught them but they didn't learn', is replaced by 'I tried this approach and it did not seem to work with some students . . . how could I do it differently?' A clear voice urging this sort of approach comes from Hermida (2009):

> So, instead of pushing non-traditional students to adopt . . . mainstream academic skills, disciplinary perspectives, and thought processes, we

> should open our classroom doors to teaching disciplinary content and academic skills from a wide array of diverse traditions so that every single—mainstream and minority—student will feel included. This will prepare both mainstream and minority students to succeed as interculturally knowledgeable citizens in a globalized world.
>
> (Hermida 2009)

While Hermida's comment does not refer specifically to engineering as a field of study, its message is certainly consistent with the message presented in this book. At one level 'doing it differently' is at the core of making the curriculum more gender inclusive: the central point of this book. We have assembled evidence that attests to the fact that some students become increasingly disengaged in engineering classrooms. Our hope in producing this work is to assist engineering educators to embrace a more inclusive approach to teaching and learning so that the poor retention rates currently seen with so many students (both female and male, traditional and non-traditional) are improved.

GENDER INCLUSIVE CURRICULUM IN PRACTICE

Chapters 5, 6 and 7 form the central core material of the book. These chapters are intended to encourage faculty to do the curriculum differently, firstly in terms of its design, secondly through examples of practice and thirdly through embedding the inclusive orientation in the course approval structures of universities and industry accreditation committees. At no time do we anticipate that doing this will be an easy task. For one thing some members of the community of engineering educators are perhaps not as aware of the issues around gender and learning as are those who have read this far in the book, and are thus yet to be convinced of the need for change. We hope that the material presented here helps to progress the understanding that there are aspects of the ways in which the engineering curriculum is taught, along with features of the content that collude in maintaining the area as a male dominated discipline. Those engineering educators who are aware of the issues for women in engineering have worked on many of the campaigns and strategies that have attempted to improve the situation. While much of their work was focussed on issues beyond the content of the curriculum, some of it was channelled into the design of courses so that they deliberately invoke gender issues in order to make engineering education gender inclusive. While it is good to see some of this happening already, it can also sometimes provoke a backlash effect as seen earlier in the quotation from one of the students. The solution is not to avoid gender inclusive modifications to the curriculum but rather to involve the class in discussion and promote understanding of why the profession stands to gain if engineers come from a much more diverse range of people than has traditionally been the case.

The search for exemplars of inclusive teaching was always considered key to the effective communication of the book's central message. We have maintained that teaching is an interpersonal encounter, and our examples comprise attempts to display personal involvement with the topics. They call on teachers and students to be actively present in the classrooms (whether they be face-to-face or virtual environments over the web) so that through their actions and vocal involvement they serve to create group coherence and interpersonal awareness. Some of this coherence is all the stronger insofar as it entails an acknowledgment of difference rather than that easy assumption of being all alike. Being a learner in this situation involves a sense of oneself as active along with an increased recognition of group membership and diversity.

The exercise of collecting and recording the exemplars in writing has been particularly difficult as it does not necessarily reflect the enthusiasm evident in the initial contact, and the exemplars can sometimes look rather flat and lifeless on the printed page. Our exemplars are not prescriptive but rather offered in order to inspire each reader to think a little differently about the topic being taught. While in some cases the actual examples may be taken up and used, more probably, for the majority, it will be some derivation, some new and different way of thinking about the topic that will be played out with the class.

It seemed to us that some of the most effective exemplars came from material that related to an interest or ongoing involvement of the person giving them. Thus the faculty member who had been very involved with sustainability issues wrote about engaging students in a project that involved monitoring of greenhouse gas emissions. In such cases it is easy to imagine the enthusiasm of the presenter being reflected in the involvement of class members. The overall lesson from the exemplars lies in the potential of transposing engineering applications into features of the students' everyday lives, reflecting the old teaching dictum—move from what they know from concrete experience into the more abstract theories and generalizable rules. If some of their experiences are drawn from other places and other cultures, then all the more reason for using them in the classroom.

The chapter addressing the embedding of inclusive principles into the criteria for university approval processes and industry accreditation is particularly important in ensuring that faculty and students are made aware of the requirement for inclusivity. In the case of the increasing diversity in student populations, it is clearly in the interests of university leadership to enable a smoother and more effective student pathway from enrolment to graduation. As we have argued here, the evidence points to the beneficial outcomes for all students from the adoption of consciously inclusive teaching and learning practices. Industry leaders and managers are becoming more aware of the need for fully functioning teamwork in the engineering projects they oversee. The idea of an educational experience in which students learn about themselves and about others who are different leads to a greater potential for working together in an increasingly diverse workforce. As one consequence of globalization, engineers are being increasingly recruited to work on projects in places around the world in teams which include a range of cultures and language backgrounds. An educational experience which more effectively prepares graduates for this sort of workplace experience would be of evident benefit to employers. Early indications are that some university managers, program leaders, industry leaders and accreditation committees have little problem with the idea of a consciously inclusive engineering curriculum. We need to continue to work towards universal acceptance of these ideas.

LOOKING AHEAD

The project that gave rise to this book was much larger than any of us envisaged when we began. Each of us had worked for some time in university departments of engineering, teaching and learning and education. As we had researched aspects of engineering education relating to gender equity we wanted to collect our findings and add our voices to the calls for more attention to the demands for inclusion and for our particular interest, gender equity. We liked the idea of a cross disciplinary approach and felt that we could add substantively to what had gone before by collecting the evidence and, in particular, by generating useful examples of engineering education working inclusively.

Looking back it seems that in the course of preparing material for the book we have learned a good deal more about the whole project of inclusive education at tertiary level than we had anticipated. We soon became most impressed by the amount of work already happening around questions of gender—much more than initially envisaged. We hope that by assembling this material in the way it is presented here, others will share our admiration for the commitment of the engineering educators who have dedicated enormous amounts of time and energy to the project of enabling more women to attain a degree in engineering and to take up a role in the profession.

In undertaking this work our attention has been drawn to the ways in which we understand the processes of teaching and learning. It seems to us that this conception is key to understanding why there is a range of different approaches to engineering education. If you believe that students learn by being shown the right way, that is, the traditional apprentice model, then your teaching will largely consist of show and tell approaches. For a significant proportion of the engineering class, especially given the stringent entry requirements that operate in many places, this approach will work. For other students it will not: not because of their lack of intellectual capacity but rather as a result of their different background experiences, different culturally acquired habits of thought and different language styles (all of which are mediated by gender, even in white middle class western households). Teaching and learning practices which include the experiences of these students should lead to more positive outcomes, but not just for them. In the course of this book we have presented evidence attesting to the fact that not all those who graduate in engineering have had a smooth, painless passage from enrolment to degree. Many students, including mainstream males, are seen to lose confidence, self esteem and engagement as they progress through the year levels. We argue then that a more inclusive approach to teaching and learning in engineering would benefit all students, not just the outliers. The mainstream students, too, would perhaps find their experience more worthwhile, memorable and enjoyable if they had increased opportunity to become more involved with the range of students in the class. Doubtless they would grow in self awareness and their subsequent work as engineers would be much more richly inflected with understanding of different ways of thinking and being in the world.

What we have proposed in this book is nothing less than a different way of doing engineering education. Already in the twenty-first century, governments are proposing that more significant proportions of the population will be able to access tertiary qualifications. This move has to be seen as a response to the credentialism creep that has been occurring for the past two decades. The anticipated larger proportion of the domestic population accessing tertiary study, along with increases in numbers of international students on campus, means that the older traditions of tertiary education will have to change. Many of these traditions were elitist to be sure, but they were also undeniably sexist. Although we now live in the times of the rhetoric of equality, instances of gender discrimination continue to exist at universities and in the workplace. We have suggested that engineering education is unfortunately still one of these places—not because it is engineering but because it operates within the constraints of the larger society and because there are people who feel compelled to maintain a belief in the immutability of gender difference.

The call for engineering education to become more gender inclusive sets up an agenda for all tertiary education. The original idea of the university, the medieval groves of academe in which great minds shared their knowledge with students sitting around their feet are long gone. We need to reclaim universities for an education in which all are equally involved as communities of educational practice. Education does not happen by osmosis but rather by people attending to the ways in which they present their learning and arrange for it to be taken up, or challenged, or at least heard and discussed. Learning is so much more than memorizing and giving back what the teacher wants to hear. Real learning demands engagement with the ideas, with the problems, with the issues. In an age where the diversity of the student body is ever increasing, education must take account of the range of different backgrounds, prior experiences and cultural interpretations that such diversity brings to all teaching and learning experiences. Engineering education is no different, and its technical nature must no longer be used as an excuse to say that it is. But engineering education has the additional challenge of including one group of students, namely women, who have remained a minority in the field since it began, in spite of decades of effort to overcome this. Our challenge to you is to help solve this problem for the profession by implementing gender inclusive engineering education in your university.

Appendix A

ACCREDITATION CRITERIA AND ATTRIBUTES OF ENGINEERING GRADUATES

Accreditation requirements for professional engineering education programs of study at bachelor degree level vary around the world but generally have consistent themes. This appendix summarizes accreditation criteria for several countries and regions of the world and provides references for further information about others. Some universities also have sets of 'graduate attributes' that they require all students at their institution to develop and hence require programs and courses to be structured around these. Some typical examples are also examined. These two sets of requirements are then brought together to provide the single framework for program and curriculum development that has been used to structure some of the discussion in Chapters 5 and 6 of this book.

Accreditation Requirements for Professional Engineering at Bachelor Degree Level

A very useful guide to the accreditation bodies responsible for engineering education accreditation in a number of countries around the world and their requirements is *Accreditation.org*, a website maintained by the Institute of Electrical and Electronics Engineers Educational Activities Board (IEEE n.d.). This site has weblinks to the accreditation organizations of many countries, a reference list for research and publications related to accreditation and details of the international mutual recognition agreements that exist between countries at the level of undergraduate qualifications. The longest standing and largest of these mutual recognition agreements is the Washington Accord (International Engineering Alliance n.d.), which was initiated in 1989 by the United Kingdom. The initial signatories to the accord were Australia, Canada, Hong Kong, Republic of Ireland, New Zealand, South Africa, United Kingdom and the US. Since that time Japan, Singapore, Korea, Chinese Taipei and Malaysia have become signatories and more countries are seeking to join each year. Countries that currently have provisional status with the accord are Germany, India, Russia and Sri

Lanka. Hence the Washington Accord incorporates the English-speaking world, key countries in Asia (except mainland China) and some of continental Europe, but has no representation from Central or South America and has limited representation from Africa. However, several South American countries (for example, Brazil, Mexico, Argentina, Chile and Peru) have based their accreditation criteria on the ABET criteria of the US and/or have Memoranda of Understanding with ABET to assist them in developing their own accreditation systems.

European countries all currently have their own accreditation bodies, but there is now a European Union–wide initiative called EUR-ACE to develop a framework for the accreditation of engineering programs throughout Europe (EUR-ACE 2008). It is administered by the European Network for Accreditation of Engineering Education (ENAEE n.d.). Another key agreement in the European region is the Bologna Declaration of 1999 (European Commission 2009). This agreement, by all members of the European Union, aims to create the European Higher Education Area (EHEA), which will have comparable academic degree and quality assurance standards in all areas of study including engineering. It agreed on standard credit transfer units and standard terminology for First Cycle (Bachelor) and Second Cycle (Master) degrees. EUR-ACE has also published an extensive report that provides an overview of accreditation procedures and criteria for Engineering programs across Europe (EUR-ACE 2005).

Some examples of accreditation criteria from various countries and agreements are detailed in this section.

United States

The accrediting body is ABET Inc. The standard duration of undergraduate engineering programs in the US is four years. ABET accredits engineering programs on a six-year cycle. There are some specific criteria about technical content and staffing within different specializations of engineering, but the overall criteria for program outcomes are as follows:

> Engineering programs must demonstrate that their students attain the following outcomes:

(a) an ability to apply knowledge of mathematics, science, and engineering
(b) an ability to design and conduct experiments, as well as to analyze and interpret data
(c) an ability to design a system, component, or process to meet desired needs within realistic constraints such as economic, environmental, social, political, ethical, health and safety, manufacturability, and sustainability
(d) an ability to function on multidisciplinary teams

(e) an ability to identify, formulate, and solve engineering problems
(f) an understanding of professional and ethical responsibility
(g) an ability to communicate effectively
(h) the broad education necessary to understand the impact of engineering solutions in a global, economic, environmental, and societal context
(i) a recognition of the need for, and an ability to engage in life-long learning
(j) a knowledge of contemporary issues
(k) an ability to use the techniques, skills, and modern engineering tools necessary for engineering practice.

(ABET 2008: 2)

Additional criteria for the curriculum also specify that there must be at least one year of mathematics and science at college level, one and a half years of engineering sciences and design and a general education component. The curriculum must also culminate in a major design experience (ABET 2008).

Australia

The accrediting body is Engineers Australia. The standard duration of undergraduate engineering programs is four years and they are accredited on a five-year cycle. The accreditation criteria are framed around three criteria: the operating environment, the academic program and quality systems. The expectations for each of these criteria are interpreted through a range of performance indicators which are intended to be used as guidelines rather than prescriptions (Engineers Australia 2008a). Those performance criteria that are relevant to gender inclusive engineering education were examined in Chapter 7. With regard to the academic program, explicit outcomes are not specified, but the accreditation guidelines suggest that the Engineers Australia National Generic Competency Standards —Stage 1 Competency Standard for Professional Engineers provides an ideal framework for this. These competency standards specify the expected knowledge base, engineering abilities and professional attributes expected of a graduating student as follows:

Knowledge base:

- Knowledge of science and engineering fundamentals
- In-depth technical competence in at least one engineering discipline
- Techniques and resources
- General knowledge.

Engineering ability:

- Ability to undertake problem identification, formulation and solution
- Understanding of social, cultural, global, and environmental responsibilities and the need to employ principles of sustainable development
- Ability to utilise a systems approach to complex problems and to design and operational performance
- Proficiency in engineering design
- Ability to conduct an engineering project
- Understanding of the business environment.

Professional attributes:

- Ability to communicate effectively, with the engineering team and with the community at large
- Ability to manage information and documentation
- Capacity for creativity and innovation
- Understanding of professional and ethical responsibilities, and commitment to them
- Ability to function effectively as an individual and in multi-disciplinary and multi-cultural teams, as a team leader or manager as well as an effective team member
- Capacity for lifelong learning and professional development
- Professional attitudes.

(Engineers Australia 2009)

A guide to the expected components of an engineering program is also specified, as follows:

- Mathematics, science, engineering principles, skills and tools appropriate to the discipline of study (not less than 40%)
- Engineering design and projects (approximately 20%)
- An engineering discipline specialisation (approximately 20%)
- Integrated exposure to professional engineering practice, including management and professional ethics (approximately 10%)
- More of any of the above elements, or other elective studies (approximately 10%).

(Engineers Australia 2008b: 10)

EUR-ACE

Engineering degrees offered across Europe have historically had different durations and requirements. The Bologna Declaration agreed on a standard framework that specified a three-year degree as a First Cycle degree and a further two years as a Second Cycle. In most countries graduates can obtain engineering work after a First Cycle degree but a Second Cycle degree is considered a full professional level qualification. The criteria that have been used for comparison purposes here are those of a Second Cycle degree level. The EUR-ACE framework specifies six program outcome areas as follows:

Knowledge and Understanding:

- An in-depth knowledge and understanding of the principles of their branch of engineering
- A critical awareness of the forefront of their branch.

Engineering Analysis:

- The ability to solve problems that are unfamiliar, incompletely defined, and have competing specifications
- The ability to formulate and solve problems in new and emerging areas of their specialisation
- The ability to use their knowledge and understanding to conceptualise engineering models, systems and processes
- The ability to apply innovative methods in problem solving.

Engineering Design:

- An ability to use their knowledge and understanding to design solutions to unfamiliar problems, possibly involving other disciplines
- An ability to use creativity to develop new and original ideas and methods
- An ability to use their engineering judgement to work with complexity, technical uncertainty and incomplete information.

Investigations:

- The ability to identify, locate and obtain required data
- The ability to design and conduct analytic, modelling and experimental investigations
- The ability to critically evaluate data and draw conclusions
- The ability to investigate the application of new and emerging technologies in their branch of engineering.

Engineering Practice:

- The ability to integrate knowledge from different branches, and handle complexity
- A comprehensive understanding of applicable techniques and methods, and of their limitations

- A knowledge of the non-technical implications of engineering practice.

Transferable Skills:

- Function effectively as an individual and as a member of a team
- Use diverse methods to communicate effectively with the engineering community and with society at large
- Demonstrate awareness of the health, safety and legal issues and responsibilities of engineering practice, the impact of engineering solutions in a societal and environmental context, and commit to professional ethics, responsibilities and norms of engineering practice
- Demonstrate an awareness of project management and business practices, such as risk and change management, and understand their limitations
- Recognise the need for, and have the ability to engage in independent, life-long learning
- Function effectively as leader of a team that may be composed of different disciplines and levels
- Work and communicate effectively in national and international contexts.

(EUR-ACE 2008: 5–8)

The EUR-ACE framework specifically does not constrain a program to contain particular components or proportions of time to develop these skills, but instead specifies an overall number of credit units to be taken.

Japan

The accrediting body is the Japan Accreditation Board for Engineering Education. The standard duration of undergraduate engineering programs is four years and they are accredited on a five-year cycle. Accredited programs are required to demonstrate that their graduates have:

(a) An ability and intellectual foundation to consider issues from a global and multilateral viewpoint.
(b) Understanding of the effects and impact of engineering on society and nature, and of engineers' social responsibility (engineering ethics).
(c) Knowledge of mathematics, natural sciences and information technology and an ability to apply such knowledge.
(d) Specialized engineering knowledge in each applicable field, and an ability to apply such knowledge to provide solutions to actual problems.

(e) Design abilities to organize comprehensive solutions to societal needs by exploiting various disciplines of science, engineering and information.
(f) Japanese-language communication skills including methodical writing, verbal presentation and debate abilities, as well as basic skills for international communication.
(g) An ability to carry on learning on an independent and sustainable basis.
(h) An ability to implement and organize works systematically under given constraints.

(JABEE 2009: 2–3)

In addition it is specified that the curriculum must comprise at least 1,800 contact hours which must include:

> . . . at least 250 hours of study in the humanities and social sciences (including language studies), at least 250 hours of study in mathematics, natural sciences and information technology, and at least 900 hours of study in the field of specialization.
>
> (JABEE 2009: 3)

A Generic Set of Graduate Attributes for Engineering

The International Engineering Alliance has developed a set of internationally agreed statements of graduate attributes and professional competencies related to the Washington Accord for professional engineers (International Engineering Alliance 2009). The broad graduate attribute categories used within this document are as follows:

- knowledge of engineering sciences
- problem analysis
- design/development of solutions
- investigation
- modern tool usage
- the engineer and society
- environment and sustainability
- ethics
- individual and team work
- communication
- project management and finance
- life-long learning.

Each of these categories then has a more detailed description of the expected abilities that an engineering graduate would develop in relation to the broad area by the completion of their professional engineering degree as recognized by the Washington Accord (International Engineering Alliance 2009).

Australian universities have been at the forefront of the development of statements of graduate attributes or qualities for all of their graduates, not just engineering. These graduate qualities are then used to frame program and course development and approval processes. One of the first universities to adopt such a framework was the University of South Australia.

> A graduate of the University of South Australia:
>
> - operates effectively with and upon a body of knowledge of sufficient depth to begin professional practice
> - is prepared for life-long learning in pursuit of personal development and excellence in professional practice
> - is an effective problem solver, capable of applying logical, critical, and creative thinking to a range of problems
> - can work both autonomously and collaboratively as a professional
> - is committed to ethical action and social responsibility as a professional and citizen
> - communicates effectively in professional practice and as a member of the community
> - demonstrates international perspectives as a professional and as a citizen.
>
> (UniSA 2009b)

For the purposes of this book and the curriculum design and delivery strategies that we have suggested within it, we have condensed and simplified all of the accreditation requirements examined earlier, along with the generic graduate qualities of the University of South Australia to develop the following set of *Attributes of Engineering Graduates* (and short-hand descriptors for them) that we have used in Chapters 5 and 6 in particular.

Attributes of Engineering Graduates

> A graduate from an undergraduate, professional level engineering degree program:
>
> - has a body of knowledge of sufficient depth to begin practice in one branch of engineering and the ability to integrate that with knowledge from different branches of engineering and related professions (Body of knowledge)
> - recognizes the need for, and has the ability to undertake, lifelong learning (Lifelong learning)
> - has the ability to identify, formulate and solve engineering problems in a logical, critical and creative way (Problem solving)

- can function professionally as both an individual and as a member of a multi-disciplinary team (Autonomy and teamwork)
- understands their professional, ethical and social responsibilities (Ethics)
- communicates effectively in both a professional and a community context (Communication)
- understands the impact of engineering solutions in global, economic, environmental and societal contexts (Global contexts).

Appendix B

GLOSSARY OF TERMS

Term	Definition Used in This Book
College	This term has not been widely used in this book, except where citing North American work. In general the term *university* has been used instead.
Course	The meaning of the terms *program*, *course* and *subject* is not consistent throughout the higher education sector. The definitions adopted in this book are those used in the US and most of Australia. Hence a *course* is the individual core or elective component of a *program*. In some countries a *course* is known as a *subject* or a *module*.
Curriculum	The entire learning experience within a *program* or *course*.
Department	The administrative unit responsible for delivering an engineering program. This term is commonly used in the US and is usually discipline based, for example Department of Civil Engineering. In Australia and sometimes in the UK the commonly used equivalent term is School.
Faculty	This term has two common meanings. The first is common in North America where it is used to describe the group of people who teach a program (and undertake research in the professional area). An individual person in this group is described as member of the faculty or simply referred to as faculty. Equivalent terms used in other countries include lecturer, professor, instructor etc. In this book we have used the term faculty or faculty member to describe the person who teaches an engineering course. The second common meaning (used occasionally in this book) is that a *Faculty* is the administrative unit in a university that encompasses a number of Departments; hence a Faculty of Engineering might include several separate departments of engineering in different disciplines. Alternative terms in common use include *College* and *Division*.

continued

Term	Definition Used in This Book
K–12	Kindergarten to Year 12 education, basically all of the years of formal education that a student will undertake prior to attending university.
Primary education	The years of formal education completed before commencing secondary education or high school. In most countries students in primary education will be aged between approximately 5 and 13, and be undertaking grades 1 to 6, but this does vary slightly.
Program	The entire requirement for qualification for a formal award of a first degree or diploma. In engineering this will typically vary between three and five years' duration depending on the country and the name of the award. A *program* will be made up of a required number of *courses*. In some countries the term course is used instead of program, combined with the term *subject* or *module* instead of course.
Secondary education	The years of formal education between primary and tertiary. In some countries this will also be called *high school* education. Generally students will be aged between 13 and 18 and be undertaking grades 7 to 12.
Subject	This term has not been used in this book, instead the term course has been used (see Course).
Tertiary education	Non-compulsory education undertaken after completion of high school, either at a university or technical college of some type.
University	An institution of tertiary education that can award degrees. This term is common throughout most of the world except North America where the term *college* is used instead, although many US colleges have 'University' in their name. The term *university* has been adopted throughout this book, except where citing North American work.

Bibliography

ABET (2008) *Criteria for Accrediting Engineering Programs. Effective for evaluations during the 2009–2010 accreditation cycle.* Online. Available HTTP: <http://www.abet.org/Linked%20Documents-UPDATE/Criteria%20and%20PP/E001%2009-10%20EAC%20Criteria%2012-01-08.pdf> (accessed 5 October 2009).

———. (2005) *Issues of accreditation in higher education Vol III: Diversity*, Baltimore: ABET Inc.

ABS (Australian Bureau of Statistics) (2009) *2006 Census of Population and Housing*, ABS data available on request.

Adelman, C. (1998) *Women and Men of the Engineering Path: a model for analyzes of undergraduate careers*, Washington: U.S. Department of Education.

ADVANCE (2009) *ADVANCE web portal*. Online. Available HTTP: <http://www.portal.advance.vt.edu/> (accessed 4 October 2009).

Agogino, A.M., Newman, C., Bauer, M. and Mankoff J. (2004) 'Perceptions of the Design Process: an examination of gendered aspects of new product development', *International Journal of Engineering Education*, 20(3): 452–60.

Ahmad, A. (1998) *Women in Malaysia. Country briefing paper*, Manila: Asian Development Bank.

Alic, M. (1986) *Hypatia's Heritage: a history of women in science from antiquity to the late nineteenth century*, London: The Women's Press.

Allen, N. and Robins, R.A. (1990) 'Increasing Women's Participation In Engineering: a review of initiatives', *Multi-disciplinary Transactions of the Institution of Engineers, Australia*, 14(2): 94–101.

Allie, S. et al. (2009) 'Learning as Acquiring a Discursive Identity through Participation in a Community: improving student learning in engineering education', *European Journal of Engineering Education*, 34(4): 359–67.

Ambrose, S., Dunkle, K., Lazarus, B., Nair, I. and Harkus, D. (1997) *Journeys of Women in the Sciences: no universal constants*, Philadelphia: Temple University Press.

Anderson, L. and Gilbride, K. (2005) 'IMAGE of Engineering among Canadian High School Students', in *Proceedings of the 8th UICEE Annual Conference on Engineering Education*, Kingston, Jamaica: UNESCO, International Centre for Engineering Education.

———. (2003) 'Pre-university Outreach: encouraging students to consider engineering careers', *Global Journal of Engineering Education*, 7(1): 87–93.

APESMA (Association of Professional Engineers, Scientists and Managers, Australia) (2007) *Women in the Professions Survey Report 2007*. Online. Available HTTP: <http://www.apesma.asn.au/women/survey_report.asp> (accessed 11 September 2009).

Armstrong, J. (1995) 'Women Engineers in Tertiary Education in Australia: the structural and cultural barriers they face', in *Proceedings of the Second National Conference on Equity and Access in Higher Education*, Melbourne.

ASEE (American Society of Engineering Education) (2009) *ASEE Connections July 2009*. Online. Available HTTP: <http://www.asee.org/publications/connections/index.cfm> (accessed 14 August 2009).

———. (1994) *The Green Report—Engineering Education for a Changing World*, Washington: ASEE. Online. Available HTTP: <http://www.asee.org/resources/beyond/greenReport.cfm> (accessed 9 September 2009).

Association of American Colleges and Universities (2009) *Diversity*. Online. Available HTTP: <http://www.diversityweb.org/> (accessed 23 September 2009).

Atkin J. and Karplus, R. (1962) 'Discovery of Invention?' *Science Teacher*, 29(5): 45.

Ayre, M. and Beynon, J. (1988) *Introductory Courses to Increase Access to Engineering*, London: The Royal Society.

Ayre, M. and Nafalski, A. (2000) 'Recognising Diverse Learning Styles in Teaching and Assessment of Electronic Engineering', paper presented at the *30th ASEE/ISEE Frontiers in Education Conference*, Kansas City, October 2000.

Bacchi, C. (1999) *Women, Policy and Politics: the construction of policy problems*, London: Sage.

Bagilhole, B. and Goode, J. (1998) 'The "Gender Dimension" of Both the "Narrow" and the "Broad" Curriculum in UK Higher Education: do women lose out on both?' *Gender and Education*, 10(4): 445–58.

Bastalich, W., Franzway, S., Gill, J., Mills, J. and Sharp, R. (2007) 'Disrupting Masculinities: women engineers and engineering culture', *Australian Feminist Studies*, 22(54): 385–400.

Bauer, R. (2006) 'Degendering Science—developing a gender and science studies curriculum', in R. Bauer and H. Gotschel (eds) *Gender in Naturwissenschaften*. Ein curriculum an der Schnittstelle er Wissenschaftskulturen. Mossingen-Talheim: Talheimer Verlag, 167–182.

Beder, S. (1989) 'Towards a More Representative Engineering Education', *International Journal of Applied Engineering Education*, 5(2): 173–82.

Bem, S. (1981) 'Gender Schema Theory: a cognitive account of sex typing', *Psychological Review*, 88(4): 354–64.

Bensimon, E. (2004) 'The Diversity Scorecard: a learning approach to institutional change', *Change*, 36(1): 45–52.

BEST (2007) *BEST—Building Engineering and Science Talent*. Online. Available HTTP: <http://www.bestworkforce.org/> (accessed 24 September 2009).

Beyer, K. (1991) 'Gender, Science Anxiety and Learning Styles', in L.J. Rennie, L.H. Parker and G.M. Hildebrand (eds) *Action for Equity: the second decade. Contributions to the Sixth International GASAT Conference: Volume 1: Schooling*, Perth: Key Centre for Teaching and Research in School Science and Mathematics (Particularly for Women).

Bianchini J., Cavazos, L. and Helms, J. (2000) 'From Professional Lives to Inclusive Practice: science teachers and scientists' views of gender and ethnicity in science education', *Journal of Research into Science Teaching*, 37(6): 511–47.

Bianchini, J., Whitney, D., Breton, T. and Hilton-Browne, B. (1999) 'Toward Inclusive Science Education: university scientists' views of students, instructional practices, and the nature of science', *Science Education*, 86(1): 42–78.

Bleier, R. (1984) *Science and Gender: a critique of biology and its theories on women*, New York: Pergamon Press.

Blickenstaff, J.C. (2005) 'Women and Science Careers: leaky pipeline or gender filter?' *Gender and Education*, 17(4): 369–86.

Blum, L. (2001) 'Transforming the Culture of Computing at Carnegie Mellon', *Computing Research News*, 13(5): 2–9.

Bowen, W., Bok, D. and Burkhart, G. (1999) 'A Report Card on Diversity: lessons for business from higher education', *Harvard Business Review*, 77(1): 138–49.

Boyer, E. (1990) *Scholarship Reconsidered: priorities of the professoriate*, Princeton: Carnegie Foundation for the Advancement of Teaching.

Brainard, S.G. and Carlin, L. (1998) 'A Six-Year Longitudinal Study of Undergraduate Women in Engineering and Science', *Journal of Engineering Education*, 87(4): 17–27.

Brew, C. (1999) 'Women Returning to Study Mathematics: an epistemological journey?' paper presented at the combined *Annual AARE and NZARE Conference*, Melbourne, November 1999.

Brown, S., Rust, C. and Gibbs, G. (1994) *Strategies for Diversifying Assessment*, Oxford: Oxford Brookes University, Oxford Centre for Staff Development.

Bunting, A., Carre, A., Kaider, F., Andrews, J., Chapple, B. and Mewburn, I. (2007) 'Greenhouse Gas Reduction in Industry: a multidisciplinary approach to project-based learning', in *Proceedings of the 2007 AAEE Conference*, Melbourne, December 2007.

Bureau of Labour Market Research (1985) *The Labour Market for Professional Engineers*, Canberra: Australian Government Publishing Service.

Burrowes, G. (2006) 'Women in Engineering Forum: statistics', notes from an unpublished presentation at the *17th Annual Conference of the Australasian Association for Engineering Education*, Auckland, December 2006.

———. (2001) 'Gender Dynamics in an Engineering Classroom: engineering students' perspectives', unpublished Master's Thesis, University of Newcastle, Electrical Engineering and Computer Science.

Busch-Vishniac, I.J. and Jarosz, J.P. (2007) 'Achieving Greater Diversity through Curricular Change', in R.J. Burke and M.C. Mattis (eds) *Women and Minorities in Science, Technology, Engineering and Mathematics*. Cheltenham: Edward Elgar, 245–275.

———. (2004) 'Can Diversity in the Undergraduate Engineering Population Be Enhanced through Curricular Change?' *Journal of Women and Minorities in Science and Engineering*, 10: 255–81.

Butler Kahle, J. Parker, L., Rennie, L. and Riley, D. (1993) 'Gender Differences in Science Education: building a model', *Educational Psychologist*, 28(4): 379–404.

Cano, R., Kimmel, H., Koppel, N. and Muldrow, D. (2001) 'A First Step for Women into the Engineering Pipeline', in *Proceedings of the 31st ASEE/IEEE Frontiers in Education Conference*, Reno, Nevada.

Capobianco, B. (2006) 'Undergraduate Women Engineering Their Professional Identities', *Journal of Women and Minorities in Science and Engineering*, 12(2–3): 95–118.

Carstensen, K. and Bernhard, J. (2009) 'Student Learning in an Electric Circuit Theory Course: critical aspects and task design', *European Journal of Engineering Education*, 34(4): 393–408.

Carter, R. and Kirkup, G. (1990) *Women in Engineering: a good place to be?* London: Macmillan.

Catalyst (2004) *The Bottom Line: connecting corporate performance and gender diversity*, New York: Catalyst.

CAWMSET (Congressional Commission on the Advancement of Women and Minorities in Science, Engineering and Technology Development) (2000) *Land of Plenty: Diversity as America's Competitive Edge in Science, Engineering and Technology*. Online. Available HTTP: <http://www.nsf.gov/publications/pub_summ.jsp?ods_key=cawmset0409> (accessed 9 September 2009).

CCWESTT (Canadian Coalition of Women in Engineering, Science, Trades and Technology) (2008) % *Women in Canadian workforce*. Online. Available HTTP: <http://www.ccwestt.org/Statistics/tabid/59/ctl/Detail/mid/461/xmid/146/xmfid/11/Default.aspx> (accessed 6 September 2009).

———. (2006) *Women in SETT*. Online. Available HTTP: <http://www.ccwestt.org/Statistics/tabid/59/Default.aspx> (accessed 5 April 2008).

Chetcuti, D. (2009) 'Identifying a Gender-Inclusive Pedagogy from Maltese Teachers' Personal Practical Knowledge', *International Journal of Science Education*, 31(1): 81–99.

Chiu, L., Chiu, S. Chiu, J. and Chiu, D. (2002) 'Engineering and Technology Education for Women in the New Century', *Engineering Science and Education Journal*, 11(4): 145–52.

Clewell, B.C. and Campbell, P.B. (2002) 'Taking Stock: where we've been, where we are, where we're going', *Journal of Women and Minorities in Science and Engineering*, 8(3–4): 255–84.

Concannon, J.P. and Barrow, L.H. (2009) 'A Cross-Sectional Study of Engineering Students' Self-Efficacy by Gender, Ethnicity, Year, and Transfer Status', *Journal of Science Education and Technology*, 18(2): 163–72.

Cooney, E.M. (1991) 'An Investigation of Gender Bias in EET Laboratories', in *Proceedings of Frontiers in Education Conference*, West Lafayette, IN, 21–24 September 1991.

Cornell Chronicle (2001) *Engineering Academy's President Will Talk on Diversity, Computing Growth*, 5 April: 7.

Coyle, E., Jamieson, L. and Oakes, W. (2005) 'EPICS: Engineering Projects in Community Service', *International Journal of Engineering Education*, 21(1): 1–12.

Cronin, C., Cooper, M. and Roger, A. (1997) *Winning Women: science, engineering and technology, a positive choice, Volume 2: Participation guide*, Edinburgh: Scottish Higher Education Funding Council.

Cronin, C. and Roger, A. (1999) 'Theorizing Progress: women in science, engineering, and technology in higher education', *Journal of Research in Science Teaching*, 36(6): 637–61.

Cross, K. (1984) *Adults as Learners*, San Francisco: Jossey-Bass.

CUE (2009) *CUE Equity Model*. Online. Available HTTP: <http://cue.usc.edu/equity_model/> (accessed 24 September 2009).

Davidson, M.J. and Burke, R.J. (2004) 'Women in Management Worldwide: facts, figures and analysis—an overview', in Davidson, M.J. and Burke, R.J. (eds) *Women in Management Worldwide: facts, figures and analysis*, Farnham: Ashgate, 1–18.

Davis, C., Ginorio, A.B., Hollenshead, C.S., Lazarus, B.B., Rayman, P.M. and Associates (1996) *The Equity Equation: fostering the advancement of women in the sciences, mathematics and engineering*, San Francisco: Jossey-Bass.

Davis, C. and Rosser, S.V. (1996) 'Program and Curricular Interventions', in C. Davis et al. (eds) *The Equity Equation: fostering the advancement of women in the sciences, mathematics and engineering*, San Francisco: Jossey-Bass, 232–264.

Dengiz, B. (2008) Opening speech for *Women in Industrial Engineering Academia 2008 workshop*, Turkey, 7–11 July 2008. Online. Available HTTP: <http://wiea2008.auburn.edu> (accessed 6 September 2009).

DEST (Department of Education, Science and Training) (2009) *Higher Education Statistics Collections*, Canberra: DEST.

———. (2006) *Audit of Science, Engineering and Technology Skills*, Canberra: DEST.

———. (2004) *Analysis of Equity Groups in Higher Education 1991–2002*, Canberra: DEST.

Doherty, L. (2002) 'Gender Benders: scientific girls and literary boys', *Sydney Morning Herald*, 20 December. Online. Available HTTP: <http://www.smh.com.au/articles/2002/12/19/1040174348370.html> (accessed 9 September 2009).

Drexel University (2009) *College of Engineering*. Online. Available HTTP: <http://www.drexel.edu/coe/about_coe.asp> (accessed 24 September 2009).

Du, X.-Y. (2006) 'Gendered Practices of Constructing an Engineering Identity in a Problem-Based Learning Environment', *European Journal of Engineering Education*, 31(1): 35–42.

Dym, C.L., Agogino, A.M., Eris, O., Frey, D.D. and Leifer, L.J. (2005) 'Engineering Design Thinking, Teaching and Learning', *Journal of Engineering Education*, 94(1): 103–120.

EHEA (European Higher Education Area) (2005) *The Framework of Qualifications for the European Higher Education Area*. Online. Available HTTP: <http://www.bologna-bergen2005.no/EN/BASIC/050520_Framework_qualifications.pdf> (accessed 14 August 2009).

ENAEE (n.d.) *European Network for Accreditation of Engineering Education*. Online. Available HTTP: <http://www.enaee.eu/enaee/presentation.htm> (accessed 5 October 2009).

Engineering Council, UK (2007) *The Accreditation of Higher Education Programmes*, London: Engineering Council UK.

EngineeringExamples.org (2009) *EngineeringExamples.org* Online. Available HTTP: <http://www.engineeringexamples.org/real%20life%20examples.htm> (accessed 8 October 2009).

Engineers Australia (2009) *Assessment of Qualifications and Competencies*. Online. Available HTTP: <http://www.engineersautralia.org.au/membership.assessment/assessment_home.cfm#stage1> (accessed 6 October 2009).

———. (2008a) *Accreditation Criteria Summary*. Online. Available HTTP: <http://www.engineersaustralia.org.au/shadomx/apps/fms/fmsdownload.cfm/file_uuid=OB1C3966-ABE4-3D41-93FC-3B96534599B5&siteName=ieaust/ (accessed 5 October 2009).

———. (2008b) *Accreditation Criteria Guidelines*. Online. Available HTTP: <http://www.engineersaustralia.org.au/shadomx/apps/fms/fmsdownload.cfm?file_uuid=0B19D0FF-0BC5-BAC1-DB36-6FB8599DDE67&siteName=ieaust>(accessed 5 October 2009).

Engineers Canada (2009) *Canadian Engineers for Tomorrow: trends in engineering enrolment and degrees awarded 2003–2007*, Ottawa: Canadian Council of Professional Engineers.

EPF (n.d.) *EPF Engineering School—History and milestones*. Online. Available HTTP: <http://www.epf.fr/html_gb/ecole/historique.html> (accessed 13 September 2009).

Epstein, D. (1997) 'Boyz' Own Stories: masculinities and sexualities in schools', *Gender and Education*, 9(1): 105.

ETAN (European Technology Assessment Network on Women and Science) (2000) *Science Policies in the European Union: promoting excellence through mainstreaming gender equality*, Luxembourg: Office for Official Publications of the European Communities.

EUR-ACE (2008) *EUR-ACE Framework Standards for the Accreditation of Engineering Programmes*. Online. Available HTTP: <http://www.enaee.eu/pdf/EUR-ACE_Framework_Standards_20110209.pdf> (accessed 5 October 2009).

———. (2005) *EUR-ACE Overview: Accreditation Criteria and Procedures for Engineering Programmes in Europe*. Online. Available HTTP: <http://www.feani.org/webfeani/EUR_ACE/PrivateSection/Documents/B1_EUR-ACE_Accreditation-Engineering-Europe.pdf> (accessed 5 October 2009).

European Commission (2009) *The Bologna Process: towards the European higher education area.* Online. Available HTTP: <http://www.ec.europa.eu/education/policies/educ/bologna/bologna_en.html> (accessed 5 October 2009).

———. (2006) *She Figures 2006: women and science statistics and indicators.* Online. Available HTTP: <http://www.ec.europa.eu/research/science-society/pdf/she_figures_2006_en.pdf> (accessed 14 August 2009).

Evetts, J. (1996) *Gender and Career in Science and Engineering*, London: Taylor and Francis.

EWB (Engineers Without Borders) (2009) *EWB challenge.* Online. Available HTTP: <http://www.ewb.org.au/ewbchallenge/> (accessed 24 September 2009).

EWEC (Extraordinary Women Engineers Coalition) (2005) *Extraordinary Women Engineers Final Report.* Online. Available HTTP: <http://www.cs.kent.ac.uk/people/staff/saf/stepping-stones/materials/EWE_Needs_Assess.pdf> (accessed 14 October 2009).

Fairweather, H. (1976) 'Sex Differences in Cognition', *Cognition* 4(3): 231–80.

FEFC (UK Further Education Funding Council) (1996) *Inclusive Learning: principles and recommendations.* Coventry: Further Education Funding Council.

Felder, R.M. (1996) 'Matter of Styles', *Prism*, 6(4): 18–23.

———. (1993) 'Reaching the Second Tier: learning and teaching styles in college science education', *Journal of College Science Teaching*, 23(5): 286–290.

Felder, R.M. and Brent, R. (2005) 'Understanding Student Differences', *Journal of Engineering Education*, 94(1): 57–72.

Felder, R.M., Felder, G.N., Mauney, M., Hamrin, C.E. and Dietz, E.J. (1995) 'A Longitudinal Study of Engineering Student Performance and Retention. III. Gender Differences in Student Performance and Attitudes', *Journal of Engineering Education*, 84(2): 151–74.

Felder, R.M. and Silverman, L. (1988) 'Learning and Teaching Styles in Engineering Education', *Journal of Engineering Education*, 78(7): 674–81.

Felder, R.M. and Soloman, B.A. (n.d.) *Learning Styles and Strategies.* Online. Available HTTP: <http://www4.ncsu.edu/unity/lockers/users/f/felder/public/ILSdir/styles.htm> (accessed 24 September, 2009).

Femtec (n.d.) *Femtec*, translated online. Online. Available HTTP: <http://www.femtec.org> (accessed 6 September 2009).

Fennema, E. and Leder, G. (1990) *Mathematics and Gender: influences on teachers and students*, New York: Teachers College Press.

Foertsch, J.M., Millar, S.B., Squire, L. and Gunter, R. (1997). *Persuading Professors: a study in the dissemination of educational reform in research institutions.* Report to the NSF Education and Human Resources Directorate, Division of Research, Evaluation, and Communication, Washington DC. Madison: University of Wisconsin–Madison, LEAD Center.

Fox, M. (1998) 'Women in Science and Engineering: theory, practice and policy in programs', *Signs*, 24(1): 201–23.

Fox, M., Sonnert, G. and Nikiforova, I. (2009) 'Successful Programs for Undergraduate Women in Science and Engineering: adapting versus adopting the institutional environment', *Research in Higher Education*, 50(4): 333–53.

Fox Keller, E. (1996) *Reflections on Gender and Science*, New Haven: Yale University Press.

———. (1985) *Reflections on Gender and Science*, New Haven: Yale University Press.

———. (1983) *A Feeling for the Organism: the life and work of Barbara McClintock*, New York: Freeman.

Frehill, L. (2007) 'The ADVANCE: Institutional transformation program's impact on engineering schools', in I. Welpe, B. Reschka and J. Larkin (eds) *Gender and Engineering: strategies and possibilities*, Frankfurt: Peter Lang.

Freire, P. (1971) *Pedagogy of the Oppressed*, New York: Herder and Herder.

Frize, M. (1999) 'Canada Doubles Enrollments of Women in Engineering in a Decade', *IEEE WIE October1999 Newsletter*. Online. Available HTTP: <http://www.ieee.org/portal/cms_docs/committee/women/newsletter_files/oct99.pdf> (accessed 9 September 2009).

Fromm, E. (2003) 'The Changing Engineering Educational Paradigm', *Journal of Engineering Education*, 92(2): 113–21.

Froyd, J.E. and Ohland, M.W. (2005) 'Integrated Engineering Curricula', *Journal of Engineering Education*, 94(1): 147–64.

Galloway, P.D. (2008) *The 21st-Century Engineer: a proposal for engineering education reform*, Reston: ASCE Press.

Galton, M. (1981) 'Differential Treatment of Boy and Girl Pupils during Science Lessons', in A. Kelly (ed) *The Missing Half*, Manchester, UK: Manchester University Press, 180–191.

Gateway Engineering Education Coalition (2008) *Gateway Engineering Education Coalition*. Online. Available HTTP: <http://www.gatewaycoalition.org/index.aspx> (accessed 24 September 2009).

Geppert, L. (1995) 'The Uphill Struggle: no rose garden for women in engineering', *Spectrum*, IEEE, 32(5): 40–50.

Gibbons, M.T. (2008) *Engineering by the Numbers*. Online. Available HTTP: <http://www.asee.org/publications/profiles/upload/2008ProfileEng.pdf> (accessed 14 August 2009).

———. (2006) *Engineering by the Numbers*. Online. Available HTTP: <http://www.asee.org/publications/profiles/upload/2006ProfileEng.pdf> (accessed 5 April 2008).

Gill, J. (2004) *Beyond the Great Divide: single sex or coeducation?* Sydney: UNSW Press.

Gill, J., Mills, J.E., Franzway, S. and Sharp, R., (2008) 'Oh you must be very clever! High achieving women, professional power and the ongoing negotiation of workplace identity', *Gender and Education*, 20(3): 223–36.

Gill, J., Mills. J.E., Sharp, R. and Franzway, S. (2005) 'Education beyond Technical Competence: gender issues in the working lives of engineers', paper presented at *Global Colloquium on Engineering Education*, Sydney, September 2005.

Gill, J., Sharp, R., Mills, J.E. and Franzway, S., (2008) 'I Still Wanna Be an Engineer! Women, education and the engineering profession', *European Journal of Engineering Education*, 33(4): 391–402.

Gill, J. and Starr, K. (2001) 'Sauce for the Goose? Deconstructing the boys and education debates', *Discourse: Studies in the Cultural Politics of Education*, 23(3): 323–34.

Gilligan C. (1982) *In a Different Voice: psychological theory and women's development*, Cambridge: Harvard University Press.

Godfroy-Genin, A.-S. and Pinault, C. (2006) 'The Benefits of Comparing Grapefruits and Tangerines: a toolbox for European cross-cultural comparisons in engineering education—using this toolbox to study gendered images of engineering among students', *European Journal of Engineering Education*, 31(1): 23–33.

Goodman Research Group (2002) *Final Report of the Women's Experiences in College Engineering (WECE) project*. Cambridge, MA: Goodman Research Group. Online. Available HTTP: <http://www.grginc.com/WECE_FINAL_REPORT.pdf> (accessed 17 September 2009).

Gorard, S., Rees, G. and Salisbury, J. (1999) 'Reappraising the Apparent Underachievement of Boys at School', *Gender and Education*, 11(4): 441–54.

Greenfield, S., Peters, J., Lane, N., Rees, T. and Samuels, G. (2002) *SET Fair Report on Women in Science, Engineering and Technology*. Online. Available HTTP:

<http://extra.shu.ac.uk/nrc/section_2/publications/reports/R1182_SET_Fair_Report.pdf> (accessed 20 September 2006).

Habermas, J. (1987) *The Theory of Communicative Action*, Cambridge: Polity Press.

Hacker, S. (1989) *Pleasure, Power and Technology: some tales of gender, engineering, and the cooperative workplace*, Boston: Unwin Hyman.

Haigh, M. (2009). 'Fostering Cross-Cultural Empathy with Non-Western Curricular Structures', *Journal of Studies in International Education*, 13(2): 271.

Hall R.M. and Sandler, B.R. (1982) *The Classroom Climate: a chilly one for women?* Washington: Association of American Colleges, Project on the Status and Education of Women.

Harding, J. (1996) 'Girls' Achievement in Science and Technology—implications for pedagogy', in P. Murphy and C. Gipps (eds) *Equity in the Classroom*, London: Falmer Press/UNESCO Publishing, 111–123.

Harding, S. (1991) *Whose Science? Whose knowledge? Thinking from women's lives*, New York: Cornell University Press.

Hartman, H. and Hartman, M. (2007) 'Is Teamwork a Female-Friendly Pedagogy?' in I. Welpe, B. Reschka and J. Larkin, (eds) *Gender and Engineering: strategies and possibilities*, Frankfurt: Peter Lang, 73–92.

———. (2006) 'Leaving Engineering: lessons learned from Rowan University's College of Engineering', *Journal of Engineering Education*, 95(1): 49–61.

Hermida, J. (2009) 'Inclusive Teaching Strategies to Promote Non-traditional Student Success', *Tomorrow's Professor Mailing List*. Online posting. Available HTTP: <http://cgi.stanford.edu/~dept-ctl/cgi-bin/tomprof/postings/php> (accessed 18 September 2009).

HESA (Higher Education Statistics Agency) (2009) *PIs 2007/08: Non-continuation rates (including projected outcomes)*. Online. Available HTTP: <http://www.hesa.ac.uk/index.php?option=com_content&task=view&id=1438&Itemid=141> (accessed 29 August 2009).

Hewlett, S.A., et al. (2008) *The Athena Factor: reversing the brain drain in science, engineering, and technology*, Boston: Harvard Business Publishing.

Heywood, J. (2005) *Engineering Education*, Hoboken: John Wiley and Sons Inc.

Hobart, E., Young, S., Mills, J. and Gill, J. (2006) 'Factors that Influence Women to Enter and Complete Engineering Studies', paper presented at the *17th Annual Conference of Australasian Association for Engineering Education*, Auckland, December 2006.

Holland, B. (2007) 'Remembering Not to Forget Why There Are So Few Women in Engineering: the Women in Engineering Program UTS 1981-', in *Proceedings of the 2007 AAEE Conference*. Online. Available HTTP <http://aaee.com.au/conference/papers/2007/index07.htm> (accessed 13 September 2009).

Holt, J. and Solomon, F. (1996) 'Engineering Education—the way ahead', *Australasian Journal of Engineering Education*, 7(1): 47–60.

hooks, b. (1994) *Teaching to Transgress: education as the practice of freedom*. New York: Routledge.

Huang, G., Taddese, N. and Walter, E. (2000) *Entry and Persistence of Women and Minorities in College Science and Engineering Education. NCES 2000–601*, U.S. Department of Education, National Center for Education Statistics, Washington. Online. Available HTTP: <http://nces.ed.gov/pubSearch/pubsinfo.asp?pubid=2000601> (accessed 9 September 2009).

Hughes, G. (2001) 'Exploring the Availability of Student Scientist Identities within Curriculum Discourse: an anti-essentialist approach to gender-inclusive science', *Gender and Education*, 13(3): 275–90.

Hyde, J., Fennema, E., Ryan, M., Frost, L. and Hopp, C. (1990) 'Gender Comparisons of Mathematics Attitudes and Affect', *Psychology of Women Quarterly*, 14(3): 299–324.

Hyde, J.S., Lindberg, S.M., Linn, M.C., Ellis, A.B. and Williams, C.C. (2008) 'Supporting Online Material for Gender Similarities Characterize Math Performance', *Science*, 320: 494–5.

IEAust (Institution of Engineers Australia) (1996a) *Changing the Culture: engineering education into the future*. Canberra: IEAust.

———. (1996b) *Changing the Culture: engineering education into the future, Task Force Reports*, Canberra: IEAust.

IEEE (Institute of Electrical and Electronic Engineers) (n.d.) *Accreditation.org: IEEE Educational Activities Board*. Online. Available HTTP: <http://www.accreditation.org/index.php> (accessed 5 October 2009).

Information Office of the State Council of the People's Republic of China (1994) *The Situation of Chinese Women*. Online. Available HTTP: <http://www.china.org.cn/e-white/chinesewoman/11-6.htm> (accessed 14 August 2009).

Ingram, S. and Parker, A. (2002) 'Gender and Modes of Collaboration in an Engineering Classroom: a profile of two women on student teams', *Journal of Business and Technical Communication*, 16(1): 33–68.

International Engineering Alliance (2009) *Graduate Attributes and Professional Competencies*. Online. Available HTTP: <http://www.washingtonaccord.org/IEA-Grad-Attr[Prof-Competencies-v2.pdf> (accessed 5 October 2009).

———. (n.d.) *Washington Accord*. Online. Available HTTP: <http://www.washingtonaccord.org/Washington-Accord/> (accessed 5 October 2009).

Isaacs, B. and Tempel, P. (2001) 'Student Projects that Celebrate Engineering: a path to diversity in the profession', in *Proceedings of the 31st ASEE/IEEE Frontiers in Education Conference*, Reno, Nevada.

ISCED (International Standard Classification of Education) (1997) Online. Available HTTP: <http://www.unesco.org/education/information/nfsunesco/doc/isced_1997.htm> (accessed 14 August 2009).

JABEE (Japan Accreditation Board for Engineering Education) (2009) *Criteria for Accrediting Japanese Engineering Education Programs Leading to Bachelor's Degree Applicable in the Year 2009*. Online. Available HTTP: <http://www.jabee.org/english/OpenHomePage/Criteria_Bachelor_2009.pdf> (accessed 5 October 2009).

Jawitz, J. and Case, J. (1998) 'Investigating Differences in the Reasons Students Give for Studying Engineering', in *Proceedings of the 10th Australasian Conference on Engineering Education*, Rockhampton: Central Queensland University.

Jawitz, J., Case, J. and Tshabalala, M. (2000) 'Why NOT Engineering? The process of career choice amongst South African female students', *International Journal of Engineering Education*, 16(6): 470–5.

Johnson, E. (1984) 'Sex Differences in Problem Solving', *Journal of Educational Psychology*, 76(6): 1359–71.

Jolly, E., Campbell, P. and Perlman, L. (2004) *Engagement, Capacity and Continuity: a trilogy for student success*. Online. Available HTTP: <(http://www.campbell-kibler.com/trilogy.pdf)> (accessed 11 September 2009).

Jolly, L. (1996) 'The First Year Engineering Ethnographic Project: an overview', paper presented at the *Third Australasian Women in Engineering Forum*, Sydney, December 1996.

Jost, R. (2004) *Benchmarks for Cultural Change in Engineering Education*, Newcastle: University of Newcastle.

Kahle, J. (2002) 'Women in Science—where are we now?' in *Proceedings of Women in Science: Opportunities in a Changing Landscape*, Bryn Mawr College, October 2001.

Kanter R. (1977) *Men and Women of the Corporation*, New York: Basic Books.

Kelly, A. (1987) 'The Construction of Masculine Science', in A. Kelly (ed) *Science for Girls?* Milton Keynes: Open University Press,

Khazanet, V. (1996) 'Women in Civil Engineering and science: it's time for recognition and promotion', *Journal of Professional Issues in Engineering Education and Practice*, 122(2): 65–8.

Kilgore, D., Chachra, D., Loshbaugh, H., McCain, J. and Yashura, K. (2007) 'Creative, Contextual, and Engaged: are women the engineers of 2020?' in *Proceedings of 114th ASEE Annual Conference and Exposition*, Honolulu.

King, R. (2008) *Addressing the Supply and Quality of Engineering Graduates for the New Century*, Strawberry Hills: Carrick Institute. Online. Available HTTP: <http://www.carrickinstitute.edu.au/carrick/webdav/site/carricksite/users/siteadmin/public/Grants_DBIprojec_engineeringquality_project%20report_25march08.pdf> (accessed 9 September 2009).

Knight, M. and Cunningham, C. (2004) 'Building a Structure of Support: an inside look at the structure of women in engineering programs', *Journal of Women and Minorities in Science and Engineering*, 10(1): 1–20.

Knowles, M. (1990) *The Adult Learner: a neglected species*, Houston: Gulf Publishing.

Kondo, D.K. (1990) *Crafting Selves: power, gender, and discourses of identity in a Japanese workplace*, Chicago: University of Chicago Press.

Kramer-Koehler P., Tooney N. and Beke, D. (1995) 'The Use of Learning Style Innovations to Improve Retention', paper presented at the *Frontiers of Education Conference*, Atlanta, November 1995.

Le Grange, L., Rochford, K. and Sass, A. (1996) 'Gender Differences among Engineering and Business/Marketing Students' Rankings of Science and Technology Related Global Problems', *Australasian Journal of Engineering Education*, 7(1): 12–19.

Lewis, S. (1995) 'Chilly Courses for Women? Some engineering and science experiences', in A.M. Payne and L. Shoemark (eds) *Women, Culture and Universities: a chilly climate?* Sydney: University of Technology.

———. (1987) personal communication.

Lewis, S., Harris, R. and Cox, B. (2007) *Engineering a better workplace: A diversity guide for the engineering profession*, 2nd ed, Barton, A.C.T.: Institution of Engineers Australia.

Lintern, S. (1995) *Oh Look . . . a girl!* Internal publication, Adelaide: University of South Australia.

Litzinger, T.A., Lee, S.H., Wise, J.C. and Felder, R.M. (2007) 'A Psychometric Study of the Index of Learning Styles', *Journal of Engineering Education*, 96(4): 309–19.

Loftus, M. (2007) 'Piercings, Not Pocket Protectors', *Prism*, 16(8). Online. Available HTTP: <http://www.prism-magazine.org/apr07/feature_tufts.cfm> (accessed 11 September 2009).

Maccoby, E. and Jacklin, C.N. (1974) *The Psychology of Sex Differences*, Stanford: Stanford University Press.

McIlwee, J. and Robinson, G. (1992) *Women in Engineering: gender, power and workplace culture*, Albany: State University of New York Press.

McLean, C., Lewis, S., Copeland, J., O'Neill, B. and Lintern, S. (1997) 'Masculinity and the Culture of Engineering', *Australasian Journal of Engineering Education*, 7(2):143–56.

McLoughlin, L. (2005) 'Spotlighting: emergent gender bias in undergraduate engineering education', *Journal of Engineering Education*, 94(4), 373–81.

Malysheva, M. (1992) 'Feminism and Bolshevism: two worlds, two ideologies', in S. Rai, H. Pilkington and A. Phizacklea (eds) *Women in the Face of Change: the Soviet Union, Eastern Europe and China*, London: Routledge, 186–199.

Maricopa Community Colleges (2009) *Diversity Infusion Program*, Online. Available HTTP: <http://www.maricopa.edu/diversityinfusion/> (accessed 23 September 2009).

Marra, R.M. and Bogue, B. (2008) 'Engineering Classroom Environments: examining differences by gender and departments', in *Proceedings of the 2008 ASEE Annual Conference*, Pittsburgh.

Marra, R.M., Rodgers, K.A., Shen, D. and Bogue, B. (2009) 'Women Engineering Students and Self-Efficacy: a multi-year, multi-institution study of women engineering student self-efficacy', *Journal of Engineering Education*, 98(1): 27–38.

Martin, D. and Arendale, D. (1994) 'Supplemental Instruction: increasing achievement and retention', in *New Directions for Teaching and Learning*, No. 60, San Francisco: Jossey-Bass.

Martino, W., Lingard, R. and Mills, M. (2004) 'Issues in Boys' Education: a question of teacher threshold knowledges?' *Gender and Education*, 16(4): 435–54.

Maskell-Pretz, M. and Hopkins, W. (1997). 'Women in Engineering: toward a barrier-free work environment', *Journal of Management in Engineering*, 13(1): 32–7.

Mattis, M. (2007) 'Upstream and Downstream in the Engineering Pipeline: what's blocking US women from pursuing engineering careers?' in R. Burke and M. Mattis (eds) *Women and Minorities in Science, Technology, Engineering and Mathematics*, Cheltenham: Edward Elgar, 334–362.

Mellodge, P. (2009) *ES143: Engineering and Design*. Online. Available HTTP: <http://uhaweb.hartford.edu/mellodge/ES143.html> (accessed 7 September 2009).

Mendick, H. (2005) 'Mathematical Stories: why do more boys than girls choose to study mathematics at AS-level in England?', *British Journal of Sociology of Education* 26(2): 235–51.

Merriam, S.B. and Caffarella, R.S. (1999) *Learning in Adulthood*, San Francisco: Jossey-Bass.

Metz, S. (2007) 'Attracting the Engineers of 2020 Today', in R. Burke and M. Mattis (eds) *Women and Minorities in Science, Technology, Engineering and Mathematics*, Cheltenham: Edward Elgar, 184–209.

Mills, J.E. and Ayre, M. (2003) 'Implementing an inclusive curriculum for women in engineering education', *ASCE Journal of Professional Issues in Engineering Education and Practice*, 129(4): 203–10.

Mills, J.E., Ayre, M. and Gill, J. (2008) 'Perceptions and Understanding of Gender Inclusive Curriculum in Engineering Education', paper presented at the *36th Annual SEFI Conference*, Aalborg, Denmark, July 2008.

Mills, J.E., Bastalich, W., Franzway, S., Gill, J. and Sharp, R. (2006) 'Engineering in Australia: an uncomfortable experience for women', *Journal of Women and Minorities in Science and Engineering*, 12: 135–54.

Mills, J.E., Mehrtens, V., Smith, E.J. and Adams, V. (2008) *CREW Revisited in 2007. The Year of Women in Engineering: an update on women's progress in the Australian engineering workforce*, Canberra: Engineers Australia.

———. (2007) 'CREW Revisited in 2007. The Year of Women in Engineering', *CHEMECA*, Melbourne, September 2007.

Mills, M. (2000) 'Issues in Implementing Boys' Programme in Schools: male teachers and empowerment', *Gender and Education*, 12(2): 221–38.

Moxham, S. and Roberts, P. (1995) *Gender in the Engineering Curriculum*, Melbourne: University of Melbourne, Equal Opportunity Unit.

Murphy, P. (1993) 'Equity and Assessment', in *GASAT (Gender and Science and Technology) 7th International Conference Proceedings*, Ontario.

Murphy, T. et al. (2007) 'Achieving Parity of the Sexes at the Undergraduate Level: a study of success', *Journal of Engineering Education*, 96(3): 241–52.

Museum of Science, Boston (2004–9) *Engineering Is Elementary*. Online. Available HTTP: <http://www.mos.org/eie/index.php> (accessed 13 September, 2009).

Museum of Science, Boston (n.d.) *Engineering the Future*. Online. Available HTTP: <http://www.mos.org/etf/> (accessed 13 September, 2009).

NAE (National Academy of Engineering) (2005) *Educating the Engineer of 2020*, Washington: National Academies Press.

———. (2004) *The Engineer of 2020: Visions of Engineering in the New Century*, Washington: The National Academies Press.

Nair, I. and Majetich, S. (1995) 'Physics and Engineering in the Classroom', in Rosser, S. (ed) *Teaching the Majority: breaking the gender barrier in science, mathematics and engineering*, New York: Teachers College Press, 25–42 .

NAS (National Academy of Sciences) (2006) *To Recruit and Advance: women students and faculty in US science and engineering*. Online. Available HTTP: <http://www.nap.edu/catalog.php?record_id=11625> (accessed 13 September 2009).

National Audit Office (2007) *Staying the Course: The retention of students in higher education*, London: National Audit Office.

NCES (National Center for Education Statistics) (2000) *Entry and Persistence of Women and Minorities in College Science and Engineering Education, NCES 2000–601*, Washington: US Department of Education. Online. Available HTTP: <http://nces.ed.gov/pubSearch/pubsinfo.asp?pubid=2000601> (accessed 17 September 2009).

Nightingale P. and Sohler C. (1994) *Considering Gender*, Campbelltown, NSW: Higher Education Research and Development Society of Australia (HERDSA).

NJIT (New Jersey Institute of Technology) (2009) *Women in Engineering & Technology Initiatives FEMME Program (2009)*. Online. Available HTTP: <http://www.njit.edu/precollege/studentprograms/summer_programs_FEMME.php> (accessed 13 September 2009)

NSB (National Science Board) (2008) *Science and Engineering Indicators 2008*. Online. Available HTTP: <http://www.nsf.gov/statistics/seind08/> (accessed 14 August 2009).

NSERC (Natural Sciences and Engineering Research Council of Canada) (2009) *Chairs for Women in Science and Engineering Program*. Online. Available HTTP: <http://www.nserc-crsng.gc.ca/Professors-Professors/CFS-PCP/CWSE-CFSG_eng.asp> (accessed 16 September 2009).

NSF (National Science Foundation) (2009) *Women, Minorities and Persons with Disabilities in Science and Engineering, NSF 09–305*, Arlington: National Science Foundation.

———. (2007) *Women, Minorities and Persons with Disabilities in Science and Engineering, NSF 07–315*, Arlington: National Science Foundation.

———. (2005) *The Engineering Workforce: current state, issues, and recommendations*, Arlington: National Science Foundation.

———. (2003) *New Formulas for America's Workforce: girls in science and engineering, NSF 03–207*, Arlington: National Science Foundation.

OECD (Organisation for Economic Cooperation and Development) (2008) *Education at a Glance 2008*. Online. Available HTTP: <http://www.oecd.org/education/database> (accessed 14 August 2009).

Ohio Resource Center (2009) *The Story of SUSTAIN*. Online. Available HTTP: <http://www.ohiorc.org/for/math/sustain.aspx> (Accessed 13 September 2009).

Ohland, M.W., Sheppard, S.D., Lichtenstein, G., Eris, O., Chachra, D. and Layton, R.A. (2008) 'Persistence, Engagement, and Migration in Engineering Programs', *Journal of Engineering Education*, 97(3): 259–78.

Osborne, L.A. (2006) *Engineering Education: an exploration of the classroom climate in sophomore engineering courses*, West Lafayette: Purdue University.

Paechter, C. (1998) *Educating the Other: gender, power and schooling*, London: Falmer.

Papadopoulos, G. (2006) 'Engineering Field Must Have Diversity', *CNET News*, 5 October. Online. Available HTTP: < http://news.cnet.com/2010–1022_3–6122825.html> (accessed 20 August 2009).

Patterson, E. (ed) (2008) *Real Life Examples in Mechanics of Solids*, East Lansing: Michigan State University.

Pawley, A. (2004) 'The Feminist Engineering Classroom: a vision for future educational innovations', in *Proceedings of the 2004 ASEE Annual Conference*, Salt Lake City.

Phipps, A. (2008) *Women in Science, Engineering and Technology: three decades of UK initiatives*, Stoke on Trent: Trentham Books.

Piaget, J. (1928/1995) 'Genetic Logic and Sociology', in J. Piaget, *Sociological Studies*, New York: Routledge.

Piaget, J. and Garcia, R. (1983/1989). *Psychogenesis and the History of Science* (H. Feider, Trans.), New York: Columbia University Press.

Plumb, C. and Reis, R.M. (2007) 'Creating Change in Engineering Education', *Change*, 39(3): 22–29.

Powell, A., Bagilhole, B. and Dainty, A. (2007) 'The Good, the Bad and the Ugly: women engineering students' experiences of UK higher education', in R.J. Burke and M.C. Mattis (eds) *Women and Minorities in Science, Technology, Engineering and Mathematics*, Cheltenham: Edward Elgar.

———. (2006) 'The Problem of Women's Assimilation into UK Engineering Cultures: can critical mass work?' *Equal Opportunities International*, 25(8): 688.

Purdue University (2009a) *EPICS: Purdue*. Online. Available HTTP: <http://engineering.purdue.edu/EPICS> (accessed 5 October 2009).

———. (2009b) *Grading Summary EPICS*. Online. Available HTTP: <https://engineering.purdue.edu/EPICS/Resources/Grading> (accessed 5 October 2009).

———. (2009c) *Women in Engineering Program*. Online. Available HTTP: <http://engineering.purdue.edu/WIEP> (accessed 15 September 2009).

RAE (Royal Academy of Engineering) (2007) *Educating Engineers for the 21st Century*, London: Royal Academy of Engineering. Online. Available HTTP: <http://www.raeng.org.uk/news/publications/list/mostrecent.html> (accessed 10 September 2009).

———. (2006) *Educating Engineers for the 21st Century: the industry view*. Online. Available HTTP: <http://www.raeng.org.uk/news/releases/henley/pdf/henley_report.pdf> (accessed 10 September 2009).

Ramsden, P. (2003) *Learning to Teach in Higher Education*, 2nd ed., London: Routledge Falmer.

Reed, L. (1999) 'Troubling Boys and Disturbing Discourses on Masculinity and Schooling: a feminist exploration of current debates and interventions concerning boys in school', *Gender and Education*, 11(1): 93–110.

Riley, D. (2008) *Engineering and Social Justice*, San Rafael: Morgan and Claypool.

———. (2003) 'Pedagogies of Liberation in an Engineering Thermodynamics Class', in *Proceedings of the 2003 American Society for Engineering Education (ASEE) Annual Conference*, Nashville, Tennessee.

Roberts, P. and Ayre, M.E. (2002) *Counting the Losses . . . an investigation of women's retention in the Australian engineering workforce*, Canberra: Engineers Australia.

Roberts, P. and Lewis, S. (1996) 'National Position Paper for Women in Engineering', *Changing the Culture: Engineering Education in the Future, Task Force Reports*, Canberra: Institution of Engineers, Australia.

Rogers, L. (2002) *Sexing the Brain*, New York: Columbia University Press.

Rosser, S. (2008) 'Building Two-Way Streets to Implement Policies that Work for Gender and Science', in L. Schiebinger (ed.) *Gendered Innovations in Science and Engineering*, Stanford: Stanford University Press, 182–197.

———. (2004) 'Gender Issues in Teaching Science', in S. Rose and B. Brown (eds) *Report on the 2003 Workshop on Gender Issues in the Sciences*. Online. Available HTTP: <http://www.colby.edu/~bbrown/2003Workshop.html> (accessed 15 October 2009).

———. (1998) 'Applying Feminist Theories to Women in Science Programs', *Signs*, 24(1): 171–99.

———. (1995) 'Reaching the Majority: retaining women in the pipeline', in S. Rosser (ed) *Teaching the Majority: breaking the gender barrier in science, mathematics and engineering*, New York: Teachers College Press, 1–21.

Rosser, S. and Kelly, B. (1994) 'From Hostile Exclusion to Friendly Inclusion: University of South Carolina system model project for the transformation of science and math teaching to reach women in varied campus settings', *Journal of Women and Minorities in Science and Engineering*, 1(1): 29–44.

Rowan University (2009) *Clinics*. Online. Available HTTP: <http://www.rowan.edu/colleges/engineering/programs/civilenvironmental/clinics.html> (accessed 8 October 2009).

Ryerson University (2009) *Women in Engineering*. Online. Available HTTP: <http://feas.ryerson.ca/styles/1/women_in_engineering/index.html> (accessed 13 September 2009).

Sagebiel, F. and Dahmen, J. (2006) 'Masculinities in Organizational Cultures in Engineering Education in Europe: results of the European Union project WomEng', *European Journal of Engineering Education*, 31(1): 5–14.

Schafer, S. (2006) 'A New Approach to Increasing Diversity in Engineering at the Example of Women in Engineering', *European Journal of Engineering Education*, 31(6): 661–71.

Severeins, S. and Ten Dam, G. (1994) 'Gender Differences in Learning Styles: a narrative review and quantitative meta-analysis', *Higher Education* 27: 487–501.

Seymour, E. (2002) 'Tracking the Processes of Change in US Undergraduate Education in Science, Mathematics, Engineering, and Technology,' *Science Education*, 85(6): 79–105.

Seymour, E. and Hewitt, N.M. (1997) *Talking about Leaving: why undergraduates leave the sciences*, Boulder: Westview Press.

Shelley, M. (1831/1992) *Frankenstein, or the Modern Prometheus* (J. Sith, ed), Boston: Bedford Books of St Martin's Press.

Sheppard, S., Macatangay, K., Colby, A. and Sullivan, W.M. (2009) *Educating Engineers: designing for the future of the field*, San Francisco: Jossey-Bass.

Sonnert, G. (1999) 'Women in Science and Engineering: advances, challenges, and solutions', in C.C. Selby (ed) *Women in Science and Engineering: choices for success*, New York: New York Academy of Sciences, 35–55.

Spears, J. (2008) 'Seeing Gender', *The Physics Teacher*, 46: 136–7.

Stäuble, B. (2004) 'Towards Gender Inclusiveness in the Teaching of Science and Engineering Subjects: a case study', in *Proceedings of the 27th HERDSA Annual Conference: Transforming Knowledge into Wisdom*, Miri, Sarawak.

Stonyer, H. (2002) 'Making Engineering Students—Making Women: the discursive context of engineering education', *International Journal of Engineering Education*, 18(4): 392–99.

Teese, R. and Polesel, J. (2004) *Undemocratic Schooling: equity and quality in mass secondary education in Australia*, Melbourne: Melbourne University Press.

Thom, M. (2001) *Balancing the Equation: where are the women and girls in science, engineering and technology?* New York: National Council for Research on Women.

Tomkinson, B., Warner, R. and Renfrew, A. (2002) 'Developing a Strategy for Student Retention', *International Journal of Electrical Engineering Education*, 29(2): 210–18.

Tucker, C.M. (2000) 'Promoting Success in Math among African American Female Children', *Women's Studies Quarterly*, 28(1–2): 143–57.

UK Resource Centre for Women in Science, Engineering and Technology (2009) Online. Available HTTP: <http://www.ukrc4setwomen.org/> (accessed 4 October 2009).

UKRC (UK Resource Centre for Women in Science, Engineering and Technology) (2009a) *Annual Pay—Gross (£) by SET occupation and gender in the UK, 2008*. Online. Available HTTP: <http://www.ukrc4setwomen.org/downloads/161_Mean_Gross_Annual_Pay_by_SET_occup_and_gender_2008.pdf> (accessed 6 September 2009).

———. (2009b) *SET Occupations in EU Countries by Gender Spring 2007*. Online. Available HTTP: <http://www.ukrc4setwomen.org/html/research-and-statistics/statistics/eu-comparisons-2009/> (accessed 29 August 2009).

———. (2009c) *Students in Higher Education 2009*. Online. Available HTTP: <http://www.ukrc4setwomen.org/html/research-and-statistics/statistics/students-in-higher-education-2009> (accessed 14 August 2009).

———. (2006) *University Students' Statistics*. Online. Available HTTP: <http://www.ukrc4setwomen.org/html/research-and-statistics/statistics/> (accessed 5 April 2008).

UniSA (2009a) *ATN Engineering Test*. Online. Available HTTP: <http://www.unisa.edu.au/itee/engineer/atn_test/default.asp> (accessed 13 September 2009).

———. (2009b) *Graduate Qualities*. Online. Available HTTP: <http://www.unisanet.unisa.edu.au/gradquals/> (accessed 5 October 2009).

———. (2009c) *UniSA Foundation Studies (Applied science and engineering)*. Online. Available HTTP: <http://www.unisanet.unisa.edu.au/programs/program.asp?Program=MFFS&Plan=ASE%2DMFFS&Year=2010> (accessed 13 September 2009).

UniSANews (2005) *Sparking a New Way to Learn Science*. Online. Available HTTP: <http://www.unisa.edu.au/unisanews/2005/September/main3.asp> (accessed 13 September 2009).

University of South Australia (1996–2009) *Equity Statistics*. Internal document.

———. (2008a) *Coursework Program Approval Manual*. Internal document.

———. (2008b) *Vision, Mission and Values*. Online. Available HTTP: <http://www.unisa.edu.au/about/intro/vision.asp> (accessed 20 September 2009).

———. (2000) *Policy for Development, Amendment and Approval of Programs and Course*. Internal document.

University of Western Australia (1996) *Do Male and Female Students Differ in Their Preferred Style of Learning?* Perth: Institutional Research Unit.

Urban Institute Education Policy Center (2000) *Summary Report on the Impact Study of the National Science Foundation's Program for Women and Girls*, Washington: National Science Foundation.

Vogt, C.M., Hocevar, D. and Hagedorn, L.S. (2007) 'A Social Cognitive Construct Validation: determining women's and men's success in engineering programs', *Journal of Higher Education*, 78(3): 337–64.

Vygotsky L.S. (1978) *Mind in Society*, Cambridge: MIT Press.

Waller, A.A. (2004) 'T-shirts and Ponytails: women students in engineering talk about self-presentation', in *Proceedings of the 2004 American Society for Engineering Education Annual Conference and Exposition*, Salt Lake City.

Warren, K. (1989) 'Rewriting the Future: the feminist challenge to the malestream curriculum', *Feminist Teacher*, 4(2–3): 46–52.

Wasburn, M. (2007) 'Strategic Collaboration and Retention of First-Year Technology Students: an instrumental case study of woman-friendly peer mentoring in a mixed-sex classroom', paper presented at the *Eighteenth Annual Conference of the Australasian Association for Engineering Education*, Melbourne,

December 2007. Online. Available HTTP: <http://aaee.com.au/conferences/papers/2007/index07.htm> (accessed 13 September 2009).

Wasburn, M.H., Wasburn-Moses, L. and Blackman, J. (2008) 'The P-16 Strategic Collaboration Model: a team mentoring approach', *The Educational Forum*, 72(1): 32–44.

Watford, B. (2007) 'Undergraduate Student Support Programs', in R. Burke and M. Mattis (eds) *Women and Minorities in Science, Technology, Engineering and Mathematics,* Cheltenham: Edward Elgar, 276–316.

Weasel, L., Honrado, M. and Bautista, D. (2000) 'The Forgotten Few: developing curricula on women in the physical sciences and engineering', *Women's Studies Quarterly,* 28(1–2): 251–63.

Webb, S. (1995) 'Female-Friendly Environmental Science', in S. Rosser (ed) *Teaching the Majority: breaking the gender barrier in science, mathematics and engineering,* New York: Columbia University, Teachers College Press, 193–211.

Wertheim, M. (1995) *Pythagoras' trousers: God, physics, and the gender wars,* London: Fourth Estate.

Whitten, B.L. and Burciaga, J.R. (2000) 'Feminist and Multicultural Pedagogy in Physics', *Women's Studies Quarterly,* 28(1–2): 213–35.

Whyte, J. (1986) *Girls into Science and Technology: the story of a project,* London: Routledge and Kegan Paul.

Willis, S. (1996) 'Gender Justice and the Mathematics Curriculum', in L. Parker, L.J. Rennie and B.J. Fraser (eds) *Gender, Science and the Mathematics Curriculum: shortening the shadow*, Dordrecht: Kluwer Academic Publishers, 41–52.

WISET (Women in Science Engineering and Technology Advisory Group) (1995) *Women in Science, Engineering and Technology: a discussion paper,* Canberra: Australian Government Publishing Service.

Women in Engineering Committee (1999) *Promoting Equity and Diversity—a guide to inclusive language and practice in engineering,* Canberra: Engineers Australia.

Womeng (2005) *Creating Cultures of Success for Women Engineers: synthesis report.* European Commission. Online. Available HTTP: <http://www.womeng.net/overview/Synthesis_Report.pdf> (accessed 13 September 2009).

Zywno, M.S., Gilbride, K.A. and Gudz, N. (2000) 'Innovative Outreach Programs to Attract and Retain Women in Undergraduate Engineering Programs', in *Proceedings of the 2nd Global Congress on Engineering Education*, Wismar, Germany.

Index

Courses and programs are indicated in *italics*

Made in the USA
Las Vegas, NV
19 September 2026

eb4ac17c-79c6-4610-a9ef-35de99822edfR01